The Western Hemisphere as seen from a weather satellite. Alternate bands of cloud and clear can be seen across the United States as storm systems move east. National Environmental Satellite Service, National Oceanic and Atmospheric Administration (NOAA) photo.

The American Weather Book

Also by David M. Ludlum

Social Ferment in Vermont (1938)
Early American Hurricanes (1963)
Early American Winters (1966)
Early American Winters II (1968)
Early American Tornadoes (1970)
Weather Record Book (1971)
The Country Journal New England Weather Book (1976)
The New Jersey Weather Book (1982)

The American Weather Book

David M. Ludlum

Houghton Mifflin Company Boston 1982

Library of Congress Cataloging in Publication Data

Ludlum, David McWilliams, date
The American weather book.

1. United States — Climate. I. Title.
QC983.L825 551.6973 81-23763
ISBN 0-395-32049-6 AACR2
ISBN 0-395-32122-0 (pbk.)

Printed in the United States of America

Portions of this book have appeared previously in *Weatherwise*.

To the thousands of men and women of America, who, either in carrying out their governmental duties or through personal initiative, have observed and recorded the mood and state of our atmosphere over the many years since the first settlement of the land.

Contents

Song of the Great Blizzard 1888

"THIRTEEN WERE SAVED"

OR NEBRASKA'S FEARLESS MAID.

SONG AND CHORUS

by WM. Vincent,

MISS MINNIE FREEMAN, SCHOOL TEACHER AND HEROINE

Whose Pluck and Good Judgment Exhibited during the Recent Fearful Blizzard in the Myra Valley District, Nebraska. Saved the lives of thirteen helpless little Children.

4

PUBLISHED BY

LYON & HEALY, CHICAGO.

COPYRIGHTED 1888.

[illegible]OES & QUENSEL LITH. CHICAGO.

A song of the Great Blizzard of 1888, commemorating the rescue of thirteen schoolchildren by their teacher during the blizzard. From the collection of the Nebraska State Historical Society, Lincoln, Nebraska.

Introduction

An Overview of American and World Weather

Our forty-eight conterminous United States make up an extensive piece of most valuable real estate that is endowed with an ideal climate for human activity. Our twenty-four degrees of latitudinal spread from north to south lie astride the halfway parallel from the Equator to the North Pole, guaranteeing temperatures of moderate range over most of the year. Annual mean readings over this 1000-mile zone vary from 78.2°F (25.7°C) at Key West, Florida, to 36.5°F (2.5°C) at International Falls, Minnesota, representative of the national limits. Occasionally, we experience periods of extreme tropical or polar conditions, usually of limited duration, but heating and cooling devices have been invented to protect us in our homes from these adversities.

In the other dimension, the longitudinal breadth of our country covers an entire continent, with wide oceans on each side, insuring an adequate supply of moisture on a yearly basis for the greater part of our territory. Annual precipitation normally varies from about 150 inches (381 millimeters) in the Olympic peninsula of Washington to about 2.0 in (51 mm) in the Death Valley of California. Though some areas endure serious drought annually, man's ingenuity has largely overcome this drawback and enabled human life and crops to flourish even in semidesert surroundings.

Overhead, our atmosphere lies within the wide band of westerly airstreams that circle the earth in the middle latitudes and largely control the sequence of weather in the temperate zone. At high altitudes in the core of this vast current lies the jet stream, a narrow ribbon of maximum-strength winds, undulating north and south in long waves while speeding around the globe. The jet stream provides the meeting place of warm and cold airstreams drawn from their native sources in the tropical or polar regions. Here they create a zone at the surface of the earth where thermal clashes take place — cold fronts, warm fronts, and cyclonic centers of low pressure are generated by these traveling weather disturbances, and high-pressure areas follow lows in a never-ending succession. The movement of these systems from west to east gives rise to our rapidly changing weather patterns, with quick switches from warm to cold, and from wet to dry conditions, the action that makes up the meteorological drama depicted on the daily weather map.

Our favorable "place in the sun" in the central zone of the northern hemisphere also enables us to enjoy the physical and mental stimulation of distinct changes in the seasons. At the winter solstice the sun stands about 42° above the horizon at Key West, Florida, and about 18° above at International Falls, Minnesota. Six months later, however, these angular altitudes have changed by about 47 degrees, with the sun at the zenith of 90° over the Straits of Florida and at an elevation of 65° at the Minnesota-Canada border. With its increased intensity and the longer duration of summer daylight, the solar energy received along the northern border of the United States increases from the December winter solstice to the March spring equinox in a ratio of 1 to 4, and to the June summer solstice in a ratio of 1 to 7. In addition to increasing the air temperature in summer, the energy input warms the surface of the land and is responsible for creating a variety of summer weather types, such as local showers, thunderstorms, and cloudbursts, which supply moisture for our fields and gardens when the rainfall of general storms is absent.

Our American climate has varied over the geologic ages, as has that of every other region of Planet Earth, from periods of arctic frigidity to equatorial heat, and from desert aridity to oceanic dampness. These basic climatic extremes have resulted from continental drift, the latitudinal and longitudinal sliding of

Opposite: A cumulonimbus raising its turrets over the Great Plains of eastern Colorado. National Center for Atmospheric Research (NCAR) photo.

the tectonic plates of the various continental masses along the surface of the globe, along with the attendant elevation and depression of their land surfaces. North America was once joined with Europe, and the sites of New York and London lay several hundreds of miles south of their present positions, their climates varying according to geographic location. Change, gradual but inexorable, has been the order of things in the climatic realm of the earth.

We have only indirect evidence of temperature and precipitation trends since the last glacial period and gain our knowledge from such evidence as variations of glacial advance and retreat, the arrangement of deposits in ocean bed cores and land ice, the spacing of tree rings, and the distribution of pollen layers on pond and bog bottoms. Not until the middle of the seventeenth century, with the introduction of scientific instruments designed to measure the atmospheric elements, were quantitive figures available. Daily weather observations of some continuity started in 1737 at Charleston, South Carolina, and in 1742 at Cambridge, Massachusetts. Over the next century a large mass of heterogeneous weather records was compiled, mainly through the initiative and dedication of individuals.

Though several attempts were made to establish networks of observers, not until the 1850s did the Smithsonian Institution develop a nationwide system of standardized observations. In 1870 the federal government assumed the responsibility, and through the agencies at various times of the Army Signal Service, the Weather Bureau, and the National Weather Service, full records have been maintained at many locations during the past 110 years. These records provide the basic data employed in this work.

The fifty United States now extend from the icy shores of the Arctic Ocean at Point Barrow, Alaska, to the warm waters of the tropical Pacific Ocean along the southern shore of the Island of Hawaii. Most of the data presented here, however, pertains to the forty-eight conterminous states that form a temperate belt across the middle of North America, where a rich variety of weather events occurs.

Before concentrating on our region in detail, it will be interesting to see how its climatic extremes compare with those in the rest of North America and throughout the world.

Weather Extremes of the United States Compared with North America and the World

TEMPERATURE

Maximum:

134°F (57°C)	Death Valley, Calif., July 10, 1913
136°F (58°C)	El Azizia, Libya, North Africa, Sept. 13, 1922

Minimum:

−69.7°F (−56.5°C)	Rogers Pass., Mont., Jan. 20, 1954
−79.8°F (−62.1°C)	Prospect Creek, Alaska, Jan. 23, 1971
−81.4°F (−63.0°C)	Snag, Yukon Terr., Canada, Feb. 3, 1947
−126.9°F (−88.3°C)	Vostok, Antarctica, Aug. 24, 1960

PRECIPITATION

Greatest in 24 hours:

43.00 in (1092 mm)	Alvin, Texas, July 25–26, 1979
38.70 in (983 mm)	Yankeetown, Fla., Sept. 5–6, 1950
73.62 in (1870 mm)	Cilaos La Reunion, Indian Ocean, Mar. 15–16, 1952

Greatest in one month:

71.54 in (1817 mm)	Helen Mine, Calif., Jan. 1909
88.01 in (2235.5 mm)	Swanson Bay, British Columbia, Nov. 1917
107.00 in (2718 mm)	Kuki, Maui, Hawaii, Mar. 1942
366.14 in (9230 mm)	Cherrapunji, Assam, India, July 1861

Greatest in one year:

184.56 in (4688 mm)	Wynochee Oxbow, Wash., 1931
332.29 in (8440 mm)	Mac Leod Harbor, Alaska, 1976
578.00 in (14681 mm)	Kuki, Maui, Hawaii, 1950
905.12 in (22990 mm)	Cherrapunji, Assam, India, 1861 (calendar year)
1041.78 in (26461 mm)	Cherrapunji, Assam, India, 1861 (season Aug. 1860–July 1861)

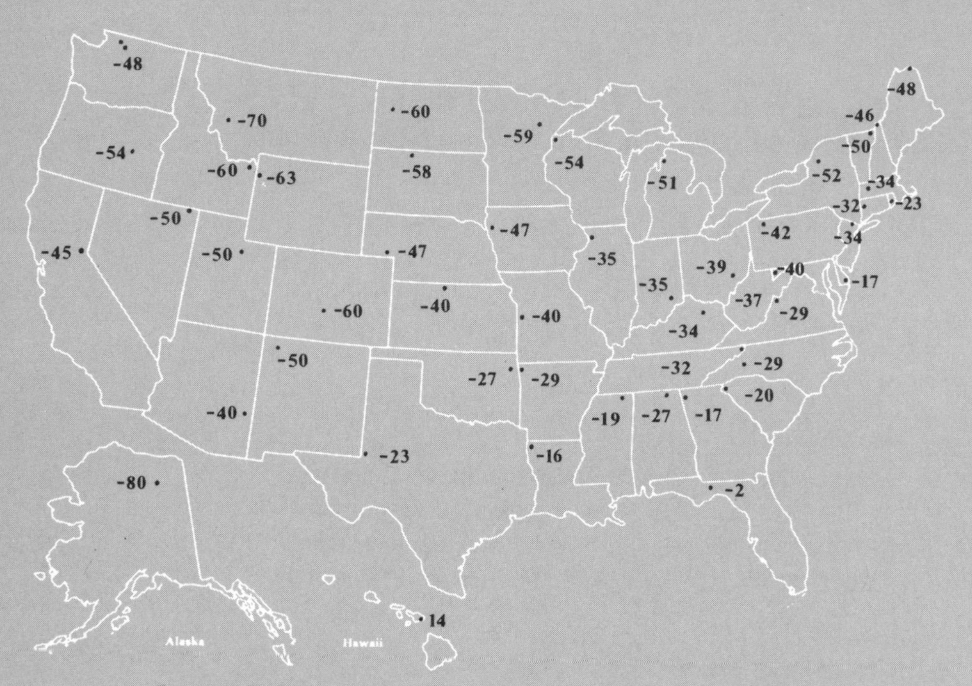

Lowest Temperatures of Record and Locations by States

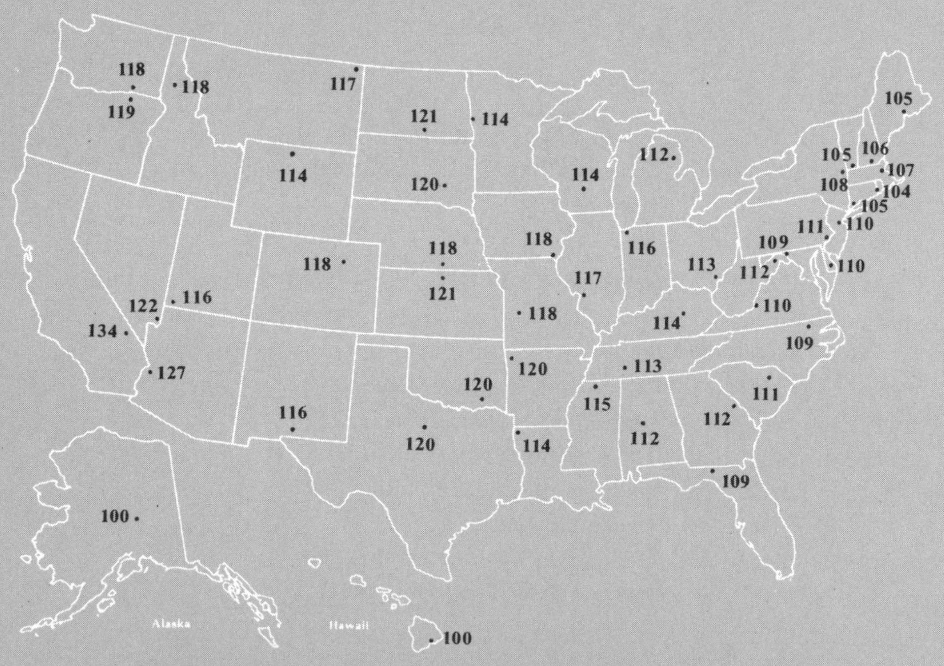

Highest Temperature of Record and Locations by States

Least in one year:

.00 in (00 mm)	Death Valley, Calif., 1919
.00 in (00 mm)	Bagdad, San Bernardino Co., Calif. Aug. 18, 1909–May 6, 1912 — total of 993 days
.00 in (00 mm)	Arica, Chile, Oct. 1903–Dec. 1917 — total of 171 months

SNOWFALL

Greatest in 24 hours:

75.8 in (193 cm)	Silver Lake, Boulder Co., Colo., Apr. 14–15, 1921

Single storm:

189.0 in (480 cm)	Mt. Shasta Ski Bowl, Calif., Feb. 13–19, 1959

Calendar month:

390.0 in (991 cm)	Tamarack, Calif., Jan. 1911

Season:

1122.0 in (2850 cm)	Paradise Ranger Station, Mt. Rainier, Wash., 1971–72

ATMOSPHERIC PRESSURE

Maximum:

31.40 in (106.3 kPa)	Helena, Mont., Jan. 9, 1962
31.43 in (106.8 kPa)	Barrow, Alaska, Jan. 3, 1970
31.53 in (106.8 kPa)	Mayo, Yukon Terr., Canada, Jan. 1, 1974
32.005 in (108.4 kPa)	Agata, Evenk N.O., Siberia (66°55′N, 93°30′E)

Minimum:

26.35 in (89.2 kPa)	Long Key, Fla., Sept. 2, 1935
25.87 in (87.6 kPa)	Typhoon June, Pacific Ocean, Nov. 19, 1975
25.69 in (87.0 kPa)	Typhoon Tip, Pacific Ocean, Oct. 12, 1979

The American Weather Book

January

Ring out the old, ring in the new,
Ring, happy bells, across the snow:
The year is going, let him go;
Ring out the false, ring in the true.

— Alfred, Lord Tennyson, In Memoriam

Jack Frost in Janiveer
Nips the nose of the nascent year.

The blackest month of all the year
Is the month of Janiveer.

A summerish January, a winterish spring
January warm, the Lord have mercy!

— Old English proverbs, collected in
Richard Inwards, Weather Lore (1898)

*Opposite: In Prospect Park, Niagara Falls, New York.
From the collection of the Library of Congress.*

"As the days lengthen, the cold strengthens," goes the familiar post-solstice proverb. January is the cold month, and its full moon is designated the Wolf Moon, according to American Indian lore. Though the direct overhead rays of the sun are moving constantly north over southern Brazil, the Northern Hemisphere continues to cool. An equilibrium between incoming solar radiation and nocturnal radiation to outer space from the snow-covered surfaces of the northlands is not reached in most places until the end of the month or in early February, some six weeks after the winter solstice has started the sun back on its journey north. Home heating requirements, as indicated by degree-day data, are at a maximum this month, not reaching the halfway mark of the winter's total fuel consumption until about January 20.

One of the interesting features of the weather this month is the more or less regular occurrence of a "January thaw." Such a deviation from the normal march of the season's temperature regime is called a "singularity" by weathermen. Old almanacs have long referred to a period toward the end of January when the snow cover melts and ice on rivers and ponds softens, giving a hint of an early spring. This period is most pronounced in the Northeast and Midwest. It usually occurs sometime between January 20 and 26, though its duration may be restricted to only a day or two. Meteorologists have made studies showing that the temperatures over the years have averaged about two degrees above the expected over this period and on particular days have greatly exceeded that figure. But it is also pointed out that the January thaw is not a fixture on the calendar; it may not come at all, as was evidenced by January 1977 and 1978, when nothing approaching a thaw put in an appearance.

Temperatures have a tendency to average above the expected from January 7 to 10, though this period is not as pronounced as later in the month nor is its appearance as frequent. There is no generally accepted reason for the occurrence of the singularities. In late January high-pressure areas or anticyclones, with their clockwise circulation of winds, tend to move more slowly than normal in some years; and this blocks the eastward movement of low-pressure areas or cyclones, with their counterclockwise winds. The southerly current set up between the two weather systems carries warmth from the South into the Midwest and Northeast and causes the temporary rise of temperatures for two or three days.

Temperature extremes for January for the conterminous forty-eight states: maximum 98°F (37°C) at Laredo, Texas, on January 17, 1954; minimum −69.7°F (−62°C) at Rogers Pass, Montana, on January 20, 1954.

In January, the controller of the transoceanic atmospheric traffic, the vast Pacific high-pressure area, lies in its southernmost position, with an west–east axis along 30°N, or the latitude of north-central Baja California and the waters north of the Hawaiian Islands. Its mean center is located about 1500 miles west of the continent, about halfway from San Diego to the Islands. This southerly position opens the way for storm systems to cross the northern Pacific Ocean from Japan and Siberia to the West Coast of North America. A secondary area of high pressure appears over the Intermountain region of Idaho, Utah, and Nevada; when this is strong, it serves to block the movement of Pacific storms inland.

South of the island chain of Alaska lies the Aleutian low-pressure area that serves as a way station for cyclonic storms crossing the North Pacific Ocean. Here they regenerate and gain energy before moving on to batter the West Coast of North America. In January the mean barometric pressure at the center of the Aleutian Low is near 29.53 inches (100 kiloPascals [kPa]).

In the Atlantic Ocean, the Azores–Bermuda high-pressure area occupies a broad band from Spain and Morocco west-southwest to the southeast coast of the United States. Its center has a mean barometric pressure of 30.20 in (102.3 kPa) and is more closely related to the Europe–North Africa landmass than to the American coast. Its affinity to the eastern Atlantic Ocean permits a low-pressure trough to form on occasion along the East Coast of the United States and serve as an important storm track for the Atlantic seaboard.

Most of the northwest North Atlantic Ocean is occupied at this season by the Iceland–Greenland low-pressure area, whose center is found off the southeast coast of Greenland in January with a mean central barometric pressure of 29.38 in (99.5 kPa). It serves as a mecca for storms crossing southern Canada or moving along the seaboard of the United States. The waters between Newfoundland, Greenland, and Iceland are notorious for severe winter storms.

Of great importance in storm activity in the northern hemisphere is the jet stream, a narrow band of high-speed winds that make up the core of the vast westerly flow circling the globe in temperate latitudes. It moves across continents and oceans in long waves with undulations north and south. The position and intensity of these waves are all-important in determining local weather since the path of the jet stream marks the convergence zone of warm southerly currents flowing from the tropics and cold northerly currents coming from the polar regions. New storms tend to generate in the vicinity of the jet stream, and it also serves to steer mature storms in their normal eastward movement around the globe.

Continuing its December behavior, the mean path of the jet stream passes around the periphery of the Pacific High and enters the North American continent over Vancouver Island of British Columbia. It then moves across the southern Prairie Provinces before dipping southeast over the western Great Lakes to the Ohio Valley, where it is often joined in wintertime by the southerly jet streaming northeast from Mexico and Texas; the combined flow moves northeast off the New Jersey coast.

The weather maps of January show a center of maximum storm frequency off Vancouver Island and the coast of Washington, where the mean path of the jet stream enters the continent. Some storm systems moving southeast from the Gulf of Alaska stall in a trough of low pressure, causing extended periods of stormy weather; eventually the low-pressure trough will move inland and cross the western mountains, spreading precipitation widely.

A second center of maximum storm frequency lies in northern Alberta where low-pressure centers, after crossing the northern Canadian Rockies, reorganize their cyclonic system of circulatory winds and develop energy before moving southeast toward the Great Lakes. Other storm centers generating in Colorado move northeast over the Great Plains, also toward the Great Lakes. The major storm track of the continent carries these disturbances from the Great Lakes down the St. Lawrence Valley to the Atlantic Ocean near Newfoundland. Another storm generation region lies in western Texas, from which disturbances of low pressure move west of the Appalachians to the Lower Lakes and then join the main storm track in the St. Lawrence Valley. When the polar front of advancing cold air dips as far south as the Gulf of Mexico, a storm-breeding area of importance to the southeast corner of the country and the Atlantic seaboard comes into action. Gulf storms cross north Florida or Georgia, then continue northeast along the coast, passing near Cape Hatteras and Cape Cod in their journey toward Newfoundland and the Iceland–Greenland Low.

January is the month when high-pressure areas achieve their greatest magnitude. These extensive areas of dense, cold air may build barometric pressure above 31.00 in (105.0 kPa) over northwest Canada, and condition air to less than −40°F (−40°C) before rushing southeast on a track that usually enters the United States west of Lake Superior. Cold waves then spread over the Great Plains and Midwest, causing some of winter's most severe weather conditions.

A second area often playing host to anticyclonic areas of high pressure lies over the Intermountain region of Idaho, Nevada, and Utah. Though not possessing the dynamic force and degree of cold of their Canadian cousins, the Intermountain highs have a marked effect on the weather west of the Rocky Mountains in blocking the eastward movement of Pacific storm centers and shunting them north over British Columbia. They may remain almost stationary for several days and are responsible for spells of stable weather that often prevail in wintertime over the western third of the nation.

Mean temperatures reach their lowest degree in January. They range from almost 0°F (−18°C) in northeast North Dakota, where Hannah has a mean of +0.7°F (−17°C), to a little over 70°F (21°C) in the Florida Keys, where Key West has 70.7°F (21°C), a difference of exactly 70°F (39°C) across the extreme latitudes of the United States. Only central and south Florida and the tip of southeast Texas average over 60°F (16°C). Across the middle of the country the line of mean freezing temperature runs from the Massachusetts coast southwest to the Appalachian Mountains of central West Virginia, then north to the vicinity of Pittsburgh, where an abrupt turn to the west-southwest carries it on to the northern Kansas border and to Denver. Some areas in southwest Arizona and Southern California average above 50°F (10°C), but most of the mountainous West exhibits a varied pattern governed by topography, altitude, and nearness to the Pacific Ocean.

There are two areas of normally heavy precipitation in January that receive more than 4.0 in (102 mm). Along the northwest Pacific Coast southward to central California, January is the height of the rainy season. Monthly amounts may average as high as 14.0 in (356 mm) in northwest Washington, while the Los Angeles area can expect about 3.0 in (76 mm). The central Gulf Coast, including a northward bulge into Tennessee and Kentucky, also has in excess of 4.0 in, though the amounts do not match those that will come in early spring.

While January is considered a between-seasons period in the Southern states, preparation of soil for spring planting is usually possible, and some seeding of oats and hardy forage crops is carried on. In the warmer portions, potatoes and hardy truck crops can usually be planted. In the northern half of the country most crops are dormant.

January

1 1864 Bitter New Year's Day in Midwest; gale-force winds, driving snow, and record low temperatures struck simultaneously; maximum at Minneapolis −25°F (−32°C), at Chicago −16°F (−27°C).

2 1910 Great flood in Nevada and Utah; 100 mi (161 km) of Salt Lake–Los Angeles railroad washed out; $7 million damage.

3 1777 Overnight freeze at Trenton enabled George Washington to flank British lines over frozen roads, cross their supply lines, and seek security in the hills of northern New Jersey after winning the short, but decisive, victory at Princeton.

4 1971 Blizzard raged from Kansas to Wisconsin; in Iowa, 27 storm-related deaths; 20 in (51 cm) of snow fell and gales mounted to 50 mi/h (80 km/h).

5 1982 In California, disaster struck San Francisco Bay and Santa Cruz areas; stalled storm front dumped 10.55 in (268 mm) at Kentfield, Marin Co., and 5.45 in (138 mm) at S.F. Airport; many damaging mudslides; 30 killed; damage over $300 million.

5 1835 Record cold morning of early nineteenth century; clear sky radiation dropped thermometers to −23°F (−31°C) on Yale campus at New Haven, to −40°F (−40°C) in Berkshire hills of Connecticut.

6 1880 Seattle's greatest snowstorm, estimated 48 in (146 cm) on ground; hundreds of barns collapsed; transportation halted; all melted soon.

7 1873 Severe blizzard struck on Great Plains; many pioneers from East, unprepared for such cold and snow, perished in Minnesota and Iowa.

8 1973 Heavy ice storm paralyzed Atlanta area for several days; business and school closings; damage estimated at $25 million.

9 1880 Second great Pacific storm within three days came ashore; barometer 28.48 in (96.4 kPa) at Olympia, Wash., 28.51 in (96.5 kPa) at Portland, Ore.; heavy snows blocked railroads for days.

10 1800 Heaviest snowfall ever known along immediate coastal plain of Southeast: Charleston 10 in (25 cm), Savannah 18 in (46 cm).

1975 Minnesota's "Storm of the Century": severe blizzard moved from south to north; barometer sank to 28.55 in (96.9 kPa) at Duluth; wind chill from −50° to −80°F (−45° to 62°C); deepest snowfall 23.5 in (60 cm); 35 storm-related deaths.

11 1918 Vast storm of blizzard proportions moved through Great Lakes and Ohio Valley; Toledo's anemometer hit 60 mi/h (97 km/h) during frontal passage, with mercury dropping from 28°F (−2°C) to −15°F (−26°C).

12 1888 Famous Blizzard of '88, western-style; sharp cold front swept from Dakotas to Texas in 24 hours; many pioneers perished when caught abroad after mild, sunny morning.

12 1982 Severe winter conditions over South; snow, sleet, and freezing rain from Texas to South Carolina, as much as 50.0 in (13 cm) deep; Atlanta, Ga., −5°F (−21°C); freeze killed vegetables in south Florida, damaged citrus in central part.

13 1888 Signal Corps thermometer dropped to −65°F (−54°C) at Fort Keogh near Miles City, Mont., long a U.S. minimum record.

14 1863 Cincinnati's greatest snowstorm began, reached depth of 20 in (51 cm) next day; modern record only 11 in (28 cm).

1979 Chicago's second heaviest snowstorm; 20.7 in (53 cm) falling in 30 hours; 29 in (74 cm) on ground established all-time depth record.

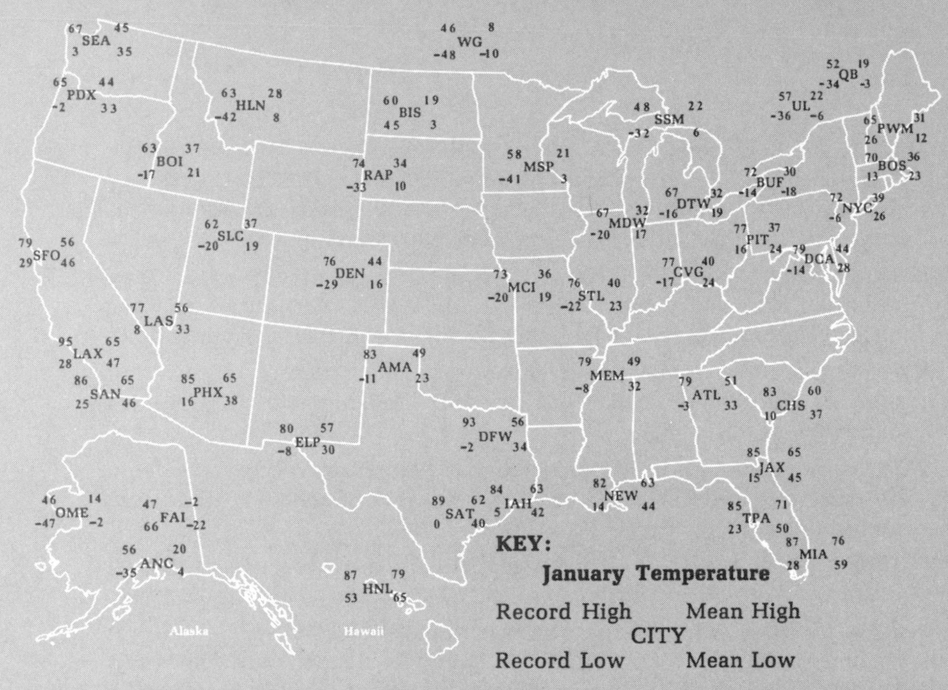

Station Designators: AMA Amarillo TX; **ANC** Anchorage AK; **ATL** Atlanta GA; **BIS** Bismarck ND; **BOI** Boise ID; **BOS** Boston MA; **BUF** Buffalo NY; **CHS** Charleston SC; **CVG** Cincinnati OH; **DCA** Washington DC; **DEN** Denver CO; **DFW** Dallas-Fort Worth TX; **DTW** Detroit MI; **ELP** El Paso TX; **FAI** Fairbanks AK; **HLN** Helena MT; **HNL** Honolulu HI; **IAH** Houston TX; **JAX** Jacksonville FL; **LAS** Las Vegas NV; **LAX** Los Angeles CA; **MCI** Kansas City MO; **MDW** Chicago IL; **MEM** Memphis TN; **MIA** Miami FL; **MSP** Minneapolis-St. Paul MN; **NEW** New Orleans LA; **NYC** New York NY; **OME** Nome AK; **PDX** Portland OR; **PHX** Phoenix AZ; **PIT** Pittsburgh PA; **PWN** Portland ME; **QB** Quebec QUE; **RAP** Rapid City SD; **SAN** San Diego CA; **SAT** San Antonio TX; **SEA** Seattle WA; **SFO** San Francisco CA; **SLC** Salt Lake City UT; **SSM** Sault Ste. Marie MI; **STL** St. Louis MO; **TPA** Tampa FL; **UL** Montreal QUE; **WG** Winnipeg MAN; **YC** Calgary ALB.

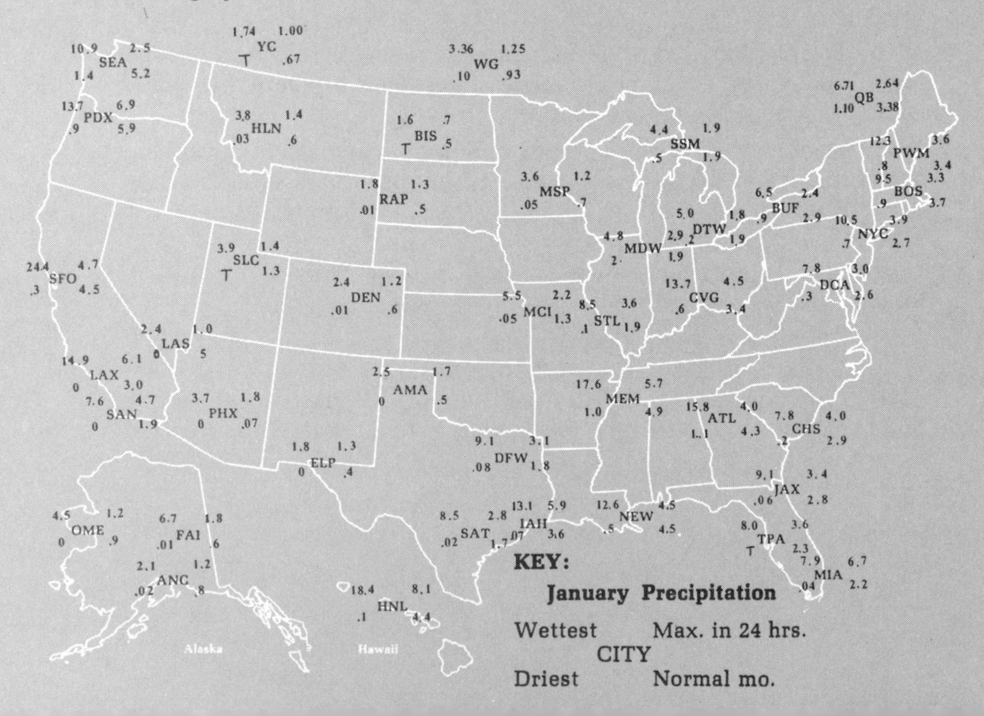

15 1932 Los Angeles' biggest snowstorm; estimated 2 in (5 cm) fell and covered ground 1 in (2.5 cm) deep at Civic Center.

16 1831 "The Great Snowstorm" raged from Georgia to Maine; 30 in (76 cm) and more fell from Pennsylvania northeast across southern New England.

17 1817 Luminous snowstorm in Vermont and Massachusetts; St. Elmo's fire appeared as static discharges on roof peaks, fence posts, and people's hats and fingers; thunderstorms prevailed over central New England.

17 1982 "Cold Sunday": Tower, Minn. −52°F (−47°C), Chicago O'Hare Airport −26°F (−32°C); zero line into central Alabama and Mississippi, Montgomery −2°F (−19°C), Jackson −5°F (−21°C); record low daylight maximums in Northeast: Buffalo −2°F (−19°C), Princeton 2°F (−17°C).

18 1857 Great cold storm swept Atlantic seaboard; temperatures near zero from Virginia north; 12-in (30-cm) snowfall general; whole gales caused structural damage on land, shipwrecks at sea; great drifts blocked transportation; Richmond cut off from Washington for seven days.

19 1810 Famous "Cold Friday" in New England; sudden overnight temperature drop of 50F (28C) degrees; gales wrecked homes; tragedy struck at Sanbornton, N.H., where three children froze to death.

1977 Snowflakes seen on extreme south of Florida peninsula; observed at Miami, Miami Beach, and Homestead at 25° 30′N.

20 1943 Strange vertical temperature antics in Black Hills of South Dakota; simultaneous readings of 52°F (11°C) at Lead, and −16°F (−27°C) at Deadwood, which are separated by only 1.5 mi (2.4 km) but have an elevation difference of 600 ft (189 m).

1937 Franklin Roosevelt's second inaugural was wettest ever; 1.77 in (50 mm) fell in Washington in 24 hours; cold, 33° (1°C) to 41°F (5°C).

21 1863 Severe coastal storm dropped heavy rains on Fredericksburg area in Virginia, disrupting Union Army offensive operation in ill-famed "Mud March."

22 1943 Black Hills thermal antics again: at Spearfish, a rise of 49F (27C) degrees in two minutes, later a fall of 58F (32C) degrees in 27 minutes; extremes ranged from −4°F (−20°C) to 54°F (12°C).

22 1982 Minneapolis had two great snowstorms in succession: 17.1 in (43 cm) on 20th and 19.9 in (51 cm) on 22–23d; these produced new 24-hour, single-storm records; also 38 in (97 cm), depth of snow on ground, and 44 in (112 cm), total depth for one month, all-time records.

23 1780 Coldest day of coldest month in history of Northeast, the famous Hard Winter; British Army thermometer dropped to −16°F

(−27°C) at New York City; harbor frozen solid for five weeks; heavy cannon dragged across ice to Staten Island; port cut off from sea supply.

24 1963 Great arctic outbreak into South, broke many records for duration of cold along Gulf Coast; Nashville set record low of −15°F (−26°C).

25 1821 Hudson River froze during coldest winter since 1780; thousands crossed on ice from New York City to New Jersey; "refreshment taverns" set up in middle of river to warm pedestrians.

26 1937 Ohio River in flood from Pittsburgh to Cairo; river 80 ft (24 m) above flood level at Cincinnati; Louisville suffered its worst inundation.

1978 Deep cyclonic storm moving north set low barometer records from Atlanta to Ontario; Sarnia, Ont., 28.21 in (95.5 kPa); hurricane-force winds, blizzard conditions in Ohio Valley and lower Lakes.

27 1967 Chicago's greatest snowstorm, 23 in (58 cm) fell in 29 hours, paralyzing city and suburbs for many days; business losses were enormous.

28 1772 Washington and Jefferson Snowstorm left 36 in (91 cm) in central and northern Virginia; described in diaries of George Washington and Thomas Jefferson.

1922 Just 150 years later came Washington, D.C.'s, deepest modern snow, the Knickerbocker Storm; 28 in (71 cm) fell, immobilizing city; crushed movie theater of that name, killing about 100 patrons.

29 1921 Great Olympic Blowdown along Washington and Oregon coasts; small, intense storm developed hurricane-force winds that funneled along mountains and felled vast expanses of Douglas fir.

30 1977 Worst blizzard of record to lee of Lakes Erie and Ontario; wind gusts 69 mi/h (111 km/h); zero visibility with whiteouts; wind chill factor −50°F (46°C); region paralyzed.

31 1966 Blizzard struck most of the Northeast; Washington, D.C., had 20 in (51 cm) on ground at end.

Cold Waves

Cold Waves — Extremes of cold in the United States result from outbreaks of polar and arctic airstreams flowing south from the interior of Alaska and the Northwest Territories of Canada. During the long northern nights the subarctic atmosphere radiates heat each night to outer space, growing colder and colder. As a result, the residual air at the surface becomes denser and heavier as it cools; this causes atmospheric pressure to build to high magnitudes.

After several days of undisturbed weather, the movement of a low-pressure system containing warm air of light density southeast along the front separating arctic and Pacific airstreams draws the cold air from the north into the rear of the developing cyclonic circulation. This greatly intensifies the storm center's energy, causing the impetuous rush south and southeast of a cold front accompanied by gale-force winds, driving snow, and rapidly falling temperatures — sometimes attaining the ferocity of a blizzard. The cold air streams to the south and east of the storm center ultimately envelop a vast portion of the United States in a cold wave.

By definition, a cold wave requires the temperature to fall by a fixed number of degrees in a specified number of hours. Different localities in the country have different criteria. For instance, at Chicago from December 15 to February 15, the temperature must fall 20F (11C) degrees in 24 hours to a minimum of 5°F (−15°C) to qualify; at Atlanta, the requirements are a drop of 20F (11C) degrees to 20°F (−7°C); and at New Orleans, a descent of only 16F (9C) degrees to 32°F (0°C) is needed.

Historic Cold Waves — There were few thermometers checking atmospheric temperatures prior to the War of Independence. All were made in England, and they were costly and difficult to transport. Probably the first in this country was brought in 1717 by Dr. Cadwallader Colden of Philadelphia. Daily thermometer readings were made for a brief period at Germantown, Pennsylvania, in 1731–32, and two series of substantial length were commenced in 1737 by Dr. John Lining of Charleston, South Carolina, and in 1742 by Professor John Winthrop of Cambridge, Massachusetts, who continued for 38 years, through the colonial period, and supplied meteorological data on the days of the stirring political events leading up to revolution. The longest series of observations, which commenced at New Haven in 1778, now supplies over 200 years of records illustrating temperature trends for the Northeast. Not until the westward movement in the first half of the nineteenth century, however, was there a sufficiently wide distribution of thermometers so that an accurate measurement of the degree of cold potential of the interior of the continent could be made.

Famous Cold Days — Three days in the eighteenth century were singled out by weather historian Ezra Stiles, president of Yale College, as memorable: 1) February 15, 1732, "reconed the coldest day in the memory of the oldest man living"; 2) February 22, 1773, "the memorable cold sabbath"; and 3) February 5, 1788, known as the "Cold Tuesday." These were days when the thermometer stayed below zero Fahrenheit (−18°C) all day with a strong northwest wind blowing and causing an extreme wind chill factor.

Cold Fridays — In New England, January 19, 1810, was long remembered for a spectacular overnight temperature drop (at Portsmouth, New Hampshire, from a balmy 41°F (5°C) in the evening of January 18 to a frigid −13°F (−25°C) in the morning). The event was marked by a family tragedy at Sanbornton, New Hampshire, when three children were frozen to death and their parents badly maimed by the cold after a chill northwest gale blew their farmhouse down and scattered their bedding and clothes across the fields. Ohio pioneers always remember February 6, 1807, as their "Cold Friday," when thermometers in central Ohio plummeted from 29°F (−2°C) to−11°F (−24°C) overnight and stayed below 0°F all daylight.

Cold Sunday in the South — Temperatures have never been lower south of the Mason-Dixon Line than they were on Sunday morning, February 8, 1835. The *Savannah Republican* reported a sunrise reading of 0°F (−18°C), and the U.S. Army's thermometer at Castle Pinckney at Charleston read 1°F (−17°C). The lowest reading reported in Florida at this time was 8°F (−13°C) on the St. Johns River near Jacksonville.

Cold January 1857 — Two arctic outbreaks in 1857 drove thermometers lower than had ever been observed in the United States. On January 18, 1857, Fort Ripley, Minnesota, reached a reading of −50°F (−46°C), and Muscatine in southeast Iowa had a record low reading of −30°F (−35°C). After the Great Cold Snowstorm of '57 swept up the Atlantic Coast on January 18–19, another cold surge of arctic air invaded the East and was soon reinforced by a second cold blast from Canada on January 22–24. A reading of −52°F (−47°C) was reported at Bath, Maine, and at Craftsbury in northern Vermont a daily mean of −28°F (−33°C) prevailed on the coldest day, January 23; this is believed to be the coldest twenty-four hours ever registered by a thermometer in the Northeast. An instrument of long reliability in downtown Boston read −5.5°F (−15°C) at 2:30 P.M. on January 23, the lowest early afternoon reading known in Boston history.

Cold New Year's Day 1864 — The outflow of cold Canadian air into the United States on the last day of 1863 brought the bitterest wind chill temperatures ever experienced in the Midwest. On January 1, 1864, St. Paul, Minnesota, averaged −29.3°F (−34.1°C) with a stiff northwest gale blowing. The Milwaukee mean figure for the day was −24.3°F (−31.3°C), and for Chicago −20.3°F (−29.1°C). The latter's minimum of −25°F (−29°C) was two degrees Fahrenheit lower than any minimum observed in 110 years of records in the Windy City until January 1982. At Louisville, Kentucky, the thermometer dropped 66.5F (39.9C) degrees in 24 hours to a low of −19.5°F (−29°C), a mark equalled only once since. The cold wave penetrated all the way to the Gulf of Mexico, and the suffering was intense with many Civil War soldiers camped in the field or housed in flimsy prison camp shacks.

Great Arctic Outbreak of 1899 — All-time marks for low readings were set from Montana south to Texas and east to the Atlantic Coast during February 11 to 14, 1899. Absolute state minimums as of that time were established in Nebraska, Texas, Louisiana, Georgia, Florida, Tennessee, Kentucky, Ohio, Pennsylvania, West Virginia, Virginia, and South Carolina. The thermometer at Tallahassee, Florida, sank to −2°F (−19°C), the lowest ever registered in the Sunshine State. Mobile reported −1°F (−18°C) and New Orleans 6.8°F (−14°C), figures still standing as their lowest of record.

Midwest — In this century severe cold periods were experienced in January 1912, January 1918, January–February 1936, January 1963, January 1977, and January 1982. Iowa, centrally located between the Mississippi and Missouri rivers, has compiled the following winter experiences: The single coldest day in most of Iowa was January 12, 1912. The coldest month of any name was January 1912, with a statewide average of 4.2°F (−15.4°C). The coldest consecutive two months were January and February 1875, with an average of 5.6°F (−14.7°C). The coldest three consecutive months were December 1874, January and February 1875, with a state average of 11.8°F (−11.2°C). But in late January and part of February 1936 all records for prolonged daily cold were broken: From January 18 to February 19, Omaha, located across the river from Council Bluffs, Iowa, averaged −2.8°F (−19.8°C) for the 33 days, or 7.7F (4.3C) degrees colder than any other period of 30 days or more since records began.

Northeast — The longest period of sustained cold ever experienced in the region from Michigan east to the North Atlantic coast came in February 1934, when temperatures for the entire month at the northernmost locations averaged close to zero Fahrenheit: New York, Stillwater Reservoir, −1.9°F (−18.8°C); Vermont, East Barnet, 0.4°F (−17.6°C); New Hampshire, Pittsburg, 0.4°F; and Maine, Ripogenus Dam, 0.0°F (−17.8°C). The frigid period began on the night of January 28–29, when a cold front dropped temperatures from well above freezing to well below zero. Thermometers reached above freezing on only one day in February at Burlington, Vermont. A severe cold outbreak on February 27–28 insured that the cold would endure through the final day and establish a record cold month. On March 2, however, thermometers rebounded to above freezing, with a week of thaw putting an end to the frigid spell. The lowest temperatures were reached on February 9 when Vanderbilt set an all-time Michigan record with −51°F (−46°C), and Stillwater Reservoir did the same in New York State with −52°F (−47°C). The lowest in New England in February 1934 was

Opposite: The winter of 1976–77 in Adams, New York. Photo by Kenneth Dewey, National Environmental Data Service, NOAA.

−42°F (−41°C), at Van Buren in northern Maine. This did not compare with the all-time low for New England of −50°F (−46°C) at Bloomfield in the Northeast Kingdom of Vermont on December 30, 1933, just six weeks earlier.

The cold extended to the seaboard as the following record lows at urban centers demonstrate: Boston −18°F (−28°C), Providence −17°F (−27°C), New York City −15°F (−26°C), and Philadelphia −11°F (−24°C).

The Coldest Winter Month: January 1977 — The "Memorable Month of January 1977" took its place among the outstanding winter months in the meteorological annals of North America. Never before has the weather made so many headlines and occupied so much front page space. This resulted not only from the severity of the weather conditions and their extreme departure from normal, but also from the critical impact on many aspects of the daily health and economic welfare of people in all parts of the nation. The total cost of the extended cold period in the East and the South and the long-continued drought in the West, which were concomitants of the January 1977 weather situation, totaled several billion dollars.

Departures from the normal January temperature exceeded 10F (6C) degrees over a vast area from the Dakotas to northeast Texas and from Florida to New Jersey. In the center along the lower Ohio Valley, departures were as great as 19F (11C) degrees at Cincinnati. The coldest temperature in the forty-eight states was −50°F (−46°C) at Lake Geneva, Wisconsin, on Jan. 9.

In the Ohio Valley it was colder than the previous coldest January, 1940; in the Southeast it was colder than January 1940; and in Florida some stations were colder than January 1940 or February 1958.

Despite the cold, snowfall amounts were generally light to moderate, except where the lake-effect mechanism was set in action by the cold winds blowing over the warm lake waters. Buffalo saw flakes in the air on every day of the month; snow in Buffalo totaled 68 in (173 cm), an all-time record; and the month ended with 38 in (97 cm) of snow on the ground, also a record. The greatest monthly total was 140 in (378 cm).

The greatest anomaly was the deep penetration of the Snow Kingdom down the Florida peninsula. For the first time in recorded history, Miamians saw snow in the air on January 19, and it reached to the latitude of the northern Keys. Freezing temperatures covered all of the mainland of the peninsula.

The Twin Cities: The Coldest Major Urban Area — The three winter months at Minneapolis–St. Paul average a mean of 16.5°F (−8.6°C), the lowest for any major urban area in the United States. Extreme readings have ranged down as low as −41°F (−41°C) in the nineteenth century St. Paul record and as low as −34°F (−37°C) in the twentieth century record at Minneapolis.

Since the first weather station was established at Fort Snelling in 1820, there have been ten months when the thermometer averaged 1°F (−17.2°C) or lower, seven occurring in the nineteenth century and three in the twentieth: Januaries in 1857, 1875, 1883, 1885, 1887, 1888, 1912, and 1917, and Februaries in 1875 and 1936. In recent years the coldest month since 1936 had been January 1963 with a mean of 2.9°F (−16°C), until January 1977 came along with a mark of 0.3°F (−18°C), to join the elite cold months.

The coldest winter of three consecutive months in the Twin Cities area during the nineteenth century came in 1874–75, when the official thermometer registered the following departures from the normal of 1940–70: December, −1.5°F (−0.8°C); January, −9.5°F (−5.3°C); and February, −16.7°F (−9.3°C).

For duration of extreme cold with the temperature remaining constantly below 0°F (−17.8°C), the winter of 1911–12 holds the prize. From December 31, 1911, to January 8, 1912, the mercury remained below 0°F (−17.8°C) for 186 consecutive hours; then it went above for a brief four hours, and there followed another period of below-zero readings of 113 consecutive hours ending on January 13. The second longest string of below-zero readings came between January 18 and 24, 1963, with 157 consecutive hours.

Record Coldest within the 48 Conterminous States

Coldest minimum:	−69.7°F (−56.5°C)	January 20, 1954	Rogers Pass, Lewis and Clark Co., Mont.

Coldest month, mean:	−19.4°F (−28.6°C)	February 1936 Turtle Lake, McLean Co., N.D.
Coldest winter, mean:	−8.4°F (−22.4°C)	December 1935–February 1936 Langdon, Cavalier Co., N.D.
Coldest year, mean:	24.5°F (−4.2°C)	1943 Mt. Washington, N.H.

Longest Duration of Severe Cold, Consecutive Days:

Langdon, Cavalier Co., N.D.

Minimum below 32°F (0°C):	176 days, October 17, 1935–April 10, 1936
Maximum below 32°F (0°C):	92 days, November 30, 1935–February 29, 1936
Minimum below 0°F (−18°C):	67 days, December 31, 1935–March 6, 1936
Maximum below 0°F (−18°C):	41 days, January 11, 1936–February 20, 1936

January 1982 Record-breaking Weather — January opened with the arctic regions of Alaska and northwest Canada encased in a bitter cold atmosphere, while the Gulf of Mexico and the Florida peninsula were covered by steaming hot-air masses. Miami hit 87°F (31°C) on New Year's Day, but Fairbanks endured a frigid −40°F (−40°C). Ominous signs of things to come.

The weather map soon went into action. Low pressure prevailed on both the Pacific and Atlantic coasts, with a strong ridge of high pressure prevailing over central sections of the country. A deep cyclonic vortex occupied the Hudson Bay region and arctic Archipelago of Canada, always a sign of mighty action to come on the weather map of North America.

Things began to happen on January 3 in central California. Marin County, along with much of the San Francisco Bay region and some localities around Monterey Bay to the south, began to slide downhill when a Pacific storm front stalled and dumped as much as 10.55 in (268 mm) of rain at Kentfield in Marin County and 6.31 in (160 mm) at the San Francisco Airport. The inevitable floods and mudslides followed, with houses sliding downhill and highways blocked. Over thirty people were killed or listed as missing.

During the very same hours on January 4–5, heavy snow blanketed eastern Wisconsin. When it was all over, the Milwaukee area was shut down; even the weather office did not open. From 8 to 16 in (20 to 41 cm) fell — the city's worst snowstorm since 1947.

All-the-while, the cold air over Canada was getting colder and colder, the atmosphere denser and denser, and the pressure was building higher and higher. Fort McMurray in northeast Alberta witnessed the thermometer at the early morning readings of: −39°, −40°, −28°, −28°, −36°, −36°, −28°, and −25°F (−29° to −40°C) on the first eight days of January. Reports of −62.7°F (−52.6°C) came from the Yukon Territory. On January 9, this frigid reservoir broke through its restraining pressure wall and spilled down into the United States, and winter at its worst followed. International Falls, Minnesota, plummeted to −36°F (−38°C) on the morning of January 9 and rose only to −24°F (−31°C) during the afternoon. Southern Florida still basked in readings of 80°F (27°C) and more.

The core of the cold air mass during the week ending on January 9 was over the American Northwest, with a departure from normal of −21°F (−12°C) in north-central Montana and a subcenter of −18°F (−10°C) over eastern Oregon. The cold pole shifted well to the southeast during the week ending January 16 to southern Illinois at −24°F (−13°C), below normal for that week.

The arctic region again restocked its reservoir of frigid air. Commencing on January 19, Fort McMurray again endured freezing mercury with the following readings: −41°, −43°, −36°, and −40°F (−38° to −42°C). The lowest reading reported at this time was −53.5°F (−47.5°C) at Uranium City, Saskatchewan. The weather map on the 19th placed a low-pressure center off the Washington coast, a second on the Nevada-Utah border, a third in southern Kansas, and a fourth on the Minnesota-Ontario border — all were producing a variety of precipitation. The principal activity came from an inactive stationary front over Kentucky and North Carolina that produced freezing rain over the Middle Atlantic States.

Heavy snow broke out over the Upper Mississippi Valley on January 20 and gave Minneapolis a record intense 16.6 in (42 cm) in 13 hours. On the 21st nearly all the country had precipitation of some kind. Moderate-to-heavy thunderstorms developed from Texas to the Carolinas as warm tropical air streamed north from the Gulf of Mexico.

Another intense winter storm developed on the southern Plains on January 22 and pushed a band of precipitation northward. Again it centered its fury on the Minneapolis area, where new records for intensity and total snowfall replaced those established only two days before: 19.9 in (51 cm) fell in 30 hours. Blizzard conditions prevailed on the Great Plains.

The weather map changed on January 24 when a surge of warm air from the central Pacific Ocean overspread the Southwest and raised temperatures well above seasonal levels. Rapid City, South Dakota, rose to 52°F (11°C), while readings were below 0°F (−18°C) in the eastern part of the state.

A Pacific storm moved inland on January 26, accompanied by light precipitation, while high pressure held the central part of the country in a rainless grip. The 27th and 28th were relatively dry. Temperatures rose above freezing along the Canadian border of Montana and Minnesota to break winter's long grip on that territory. A real January thaw prevailed west of the Mississippi River.

Winter, however, was not through. On January 29, another surge of cold air crossed the Canadian border, dropping the thermometer below 0°F (−18°C) on the morning of the 30th and to arctic depths on the 31st, when International Falls endured a minimum of −35°F (−37°C).

Meantime, another great storm was brewing over the southern Plains. On January 30, a very sharp cold front ran from the Great Lakes to Texas with strong north winds bearing arctic air southward, and equally strong southerly winds streaming north over the lower Mississippi Valley. The temperature at Memphis soared to 59°F (15°C) on the 30th, while in the same latitude west of the front Oklahoma City read 39°F (4°C) and falling rapidly.

A band of rain, sleet, freezing rain, and snow covered the middle Mississippi Valley. Over 2.00 in (51 cm) fell from Arkansas to Ohio, with locally heavier amounts at Cairo and Evansville. Only 125 miles (200 km) to the northwest in central Illinois, 22 in (56 cm) of snow descended. St. Louis had 19 in (48 cm), for its biggest snowstorm since 1890, and coming in only 1.4 in (3.6 cm) short of that. The snow band extended northeast to Michigan, where Auto City was completely paralyzed by 10 in (25 cm) of blowing snow. Indianapolis and Kokomo in Indiana received 17 in (43 cm) apiece and endured snow blockades.

The cold of January 1982 centered in the northern Great Plains and the Prairie Provinces of Canada. Minus departures from normal were as great as −15°F (−8°C) in Montana and the Dakotas. It was also very cold eastward along the Canadian border, with −10°F (−6°C) in Wisconsin and Michigan, and −11°F (−6°C) in New Hampshire. The cold this winter did not equal that of January 1977 in the Midwest and Ohio Valley, when minus departures from normal were as great as −19°F (−11°C). But January 1982 was much wetter, with frequent heavy snowstorms, blizzard conditions, severe ice storms, and unprecedented wind chill.

The Ten Days that Shook the Weather Record Book: January 1982

Jan. 9 Arctic outbreak into the Upper Midwest, propelled by a 31.10 in (105.2 kPa) anticyclone over northern Alberta; International Falls, Minn., had minimum −36°F (−38°C) and maximum −24°F (−31°C); lowest in Wisconsin, −35°F (−37°C) at Morse; extreme of −56°F (−49°C) reported in Yukon Territory of Canada.

Jan. 10 High pressure at 31.15 in (105.6 kPa) over central Saskatchewan with ridge south to Texas; zero °F line into Kansas and Missouri; lowest in nation, −37°F (−38°C) at Bemidji, Minn. In Illinois, Wisconsin, and Ohio, all-time minimum record for Chicago: O'Hare and Midway airports −26°F (−32°C), University of Chicago near lakefront, −21°F (−29°C); new records also: −23°F (−31°C) at Moline, −18°F (−28°C) at Peoria; Milwaukee at −25°F (−32°C) equaled all-time record; game time at Bengals–Forty Niners playoff game at Cincinnati was −9°F (−23°C); 41 weather stations reported new date minimum records.

Jan. 11 High centered over eastern Oklahoma and Texas; zero line into central Alabama and Georgia: Atlanta −5°F (−21°C), Birmingham −2°F (−19°C); freezing line into central Florida: Pensacola 8°F (−13°C), Orlando 32°F (0°C); 34 stations reported new date minimum records; Buffalo buried under 28 in (71 cm) snowburst; 25.3 in (64 cm) in 24 hours set new record.

Jan. 12 Freezing into south Florida — vegetables killed: Homestead 29°F (−2°C); citrus damaged: Orlando 23°F (−5°C). Gulf low-pressure area spreads snow and freezing rain over central and north Gulf States; generally 5 in (13 cm) from north Louisiana to north Florida; Atlanta paralyzed by snow and freezing rain.

Jan. 13 Atlantic Coast low spreads 6–8 in (15–20 cm) of snow in Middle Atlantic and New England. Temperature rebounds in Florida to 82°F (28°C) at Miami, to 80°F (27°C) at Fort Myers.

Jan. 14 Cold day in upper Mississippi Valley: Minneapolis, minimum −21°F (−29°C), maximum −3°F (−19°C). Heavy rain in north Florida, Jacksonville 1.45 in (37 mm) from low center passing just to south.

Jan. 15 Deep low off Nova Scotia at 28.70 in (97.2 kPa) and large high over British Columbia and Alberta at 30.84 in (104.4 kPa). Heavy lake-effect snows along Great Lakes. Zero morning again in Ohio, Indiana, and Illinois. Freeze again in central and north Florida.

Jan. 16 Second severe arctic outbreak; vast anticyclone from British Columbia to Louisiana, centered over Kansas at 30.50 in (103.6 kPa); zero line into Texas, Amarillo dropped from 55°F (13°C) to −1°F (−18°C), Oklahoma City from 62°F (17°C) to 10°F (−12°C). Continued heavy snow squalls over Great Lakes.

Jan. 17 — "Cold Sunday" High pressure covered all east of the Rockies, centered over Kentucky to Arkansas at 30.47 in (103.2 kPa). Lowest in nation: −52°F (−47°C) near Tower, Minn.; International Falls −45°F (−43°C), Chicago O'Hare −25°F (−32°C). Very cold daylight hours in East: Buffalo −2°F (−19°C) maximum, Princeton, N.J., 2°F (−17°C) at 2:00 P.M. Minimums: Buffalo −16°F (−27°C) for new January record; Milwaukee −26°F (−32°C), coldest in 111 years; Washington, D.C., −5°F (−21°C), coldest since 1934. Zero again in South: Montgomery, Ala., −2°F (−19°C), Jackson, Miss., −5°F (21°C). Florida escapes freeze. Canyon winds in Colorado hit 135 mi/h (217 km/h) in Lyons County; extensive structural damage around Boulder.

Jan. 18 Another cold, cold morning in Northeast: Princeton, N.J., −9°F (−23°C), Philadelphia −4°F (−20°C), Bridgehampton on Long Island −10°F (−23°C), close to all-time record.

February

Good-morrow, Benedict; why, what's
the matter,
That you have such a February face,
So full of frost, of storm, of cloudiness?
— Shakespeare, Much Ado About Nothing

Warm February, bad hay crop;
Cold February, good hay crop.

When gnats dance in February,
The husbandman becomes a beggar.
— Richard Inwards, Weather Lore (1898)

Let it snow, let it snow, let it snow" provides the theme for February. One cannot do much about it. The season, as might be expected, brings us the full Snow Moon. Almost everywhere in the United States more snow falls per day in February than in any other month. The proper combination of below-freezing temperatures and precipitable moisture is present in some years to produce snowfall even at extreme southern latitudes. Brownsville, near the southernmost point of Texas at 25°54′N, and south Florida, as far south as the beginning of the Keys at 25°30′N, have experienced the fall of snowflakes during the period from mid-January to mid-February. On two February occasions, in 1895 and 1899, the entire Gulf Coast from southeast Texas around to central Florida witnessed substantial snowfalls, and as recently as 1973, a zone across Georgia and South Carolina was buried under deeper snowfalls than many northern cities have ever measured.

The outstanding annual date in weather folklore comes on February 2, or forty days after the celebration date of the birth of Christ. In church circles this is the Feast of the Purification of the Virgin and also Candlemas Day, when the candles to be used for the rest of the year are blessed. The ceremony was established in A.D. 542 by the edict of Emperor Justinian I. Somehow in Western Europe it came to be considered the midpoint of winter, when people searched for some way of telling what the other half of the winter might hold in store. Germans depended on the behavior of a hibernating bear or badger to give a sign if winter was over, or whether it would continue until the equinox. When they came to America, the role was transferred to the ground hog or woodchuck, and much was made in the German settlements of its winter behavioral habits. Now the day is a sort of holiday in Punxsutawney in west-central Pennsylvania. Under the aegis of the local chamber of commerce, the news media gather on the morning of February 2 every year to see whether the ground hog when emerging from its hole can see a shadow. Scientific investigations have not shown the ground hog to possess a sixth sense or more prophetic power than other animals, but the legend of its prowess continues from year to year undiminished among a public who are fond of consulting clairvoyants in weather as well as in other matters.

Snowballing and snow-rolling at corner of Market and Post streets, San Francisco, during three-inch snowstorm on December 31, 1882. Bancroft Library (University of California, Berkeley) photo.

The Pacific High attains its most southerly position in February with the mean position of the center along latitude 29°N, well off the north Baja California coast. This location permits Pacific Coast storms to generate at relatively low latitudes, making February and March the stormiest period of the year in Southern California. A subcenter of high pressure is found over Idaho and Utah, though this is now declining in strength and influence from its early winter importance. The Aleutian Low has expanded, with its center now lying astride the larger islands, making for a stormy time on the mainland.

In the Atlantic Ocean, the main portion of the Azores–Bermuda High lies off the northwest African coast, with a long arm reaching west to southern Florida, creating stable weather conditions over the vast oceanic area. The Iceland–Greenland Low occupies the southeast corner of Greenland and adjacent waters; it continues to attract storm centers moving off the North American continent and to reenergize them for a transatlantic crossing to Europe.

The jet stream now crosses the Pacific Ocean at its most southerly position, close to latitude 35°N. Upon approaching the North American continent, it dips southeast, skirting the arm of the Pacific High over Baja California; then a northeast trend carries it to Tennessee for a junction with a smaller jet coming from the northwest. Together, they move off the coast in the vicinity of the Virginia Capes, resulting in a stormy February along the Middle Atlantic and southern New England coasts.

With the westerly airstreams occasionally reaching low latitudes, the Gulf of Mexico becomes a storm-generation area in February. Most of the snowstorms that affect the Gulf States and Florida take place in February. Many disturbances generating in the Gulf of Mexico move north along either flank of the Appalachians and cause the famous northeasters that make February the snowiest month from the Great Lakes to New England. The southerly position of the westerly flow across the Pacific Ocean stimulates a secondary storm track across California and into the Intermountain region of Nevada and Utah, where storm frequency in February is greater than in the Alberta storm factory to the north in Canada.

The principal tracks of polar anticyclones are about the same as in January, though there is some southward displacement evident. The southeast flow of polar air from northwest Canada is still a principal feature of winter atmospheric action. The snow-covered tundra and frozen surface of Hudson Bay serve as chill air-conditioning areas for polar and arctic airstreams destined to descend on the Midwest and Eastern states.

Temperature extremes in February: maximum 105°F (41°C) at Montezuma, Arizona, on February 3, 1963; minimum −66°F (−54°C) at West Yellowstone, Montana, February 9, 1933.

Mean temperatures begin to respond to the increasing solar elevation. The coldest station again is Hannah, North Dakota, but the mean has risen five full degrees to 5.7°F (−15°C). Southern

Florida remains the warmest area, where Key West shows an increase of 0.9F (0.5C) degrees over January. The 60°F (16°C) mean isotherm makes small advances northward in Texas and Florida, and the 50°F (10°C) isotherm encompasses about half the Gulf States and half of Texas. The 30°F (−1°C) line edges northward but only slightly. The 60°F (16°C) area in the Southwest expands considerably, extending north into the Great Valley of California.

The general February precipitation pattern resembles that of January. The Gulf States area shows an increase in size and amount falling, while rain on the Pacific Coast decreases from the maximums of January — except in Southern California, where February is usually a little wetter than the preceding month. The Great Plains receive little Gulf of Mexico moisture at this season and remain very dry.

Sunshine increases markedly over January: in the Great Lakes area (Detroit by 33 percent) and in the Northwest (Portland, Oregon, by 50 percent).

Tornado activity shows a definite increase, being 40 percent more frequent in February than in January. As many as 83 tornadoes were recorded in February 1971. An extreme outbreak occurred on February 19, 1884, when some 60 tornadoes killed at least 800 people in Alabama, Georgia, South Carolina, and North Carolina. The center of maximum tornado frequency remains in its winter locale in the Gulf States.

During February, vegetation continues practically dormant in the northern half of the country, except on the Pacific Coast. In the southern sections hardy truck crops are planted and grown in the open, especially in south Florida and the lower Rio Grande Valley. The seeding of early truck crops is accomplished in the southern tier of states; the planting of corn normally begins in southernmost Texas early in the month.

February

1 1920 Anticyclone of great magnitude over Northeast: Northfield, Vt., barometer 31.14 in (105.4 kPa); Portland, Me., 31.09 in (105.3 kPa), highest ever in the United States at sea level; four-day snow-and-sleet storm paralyzed region, February 4 to 7.

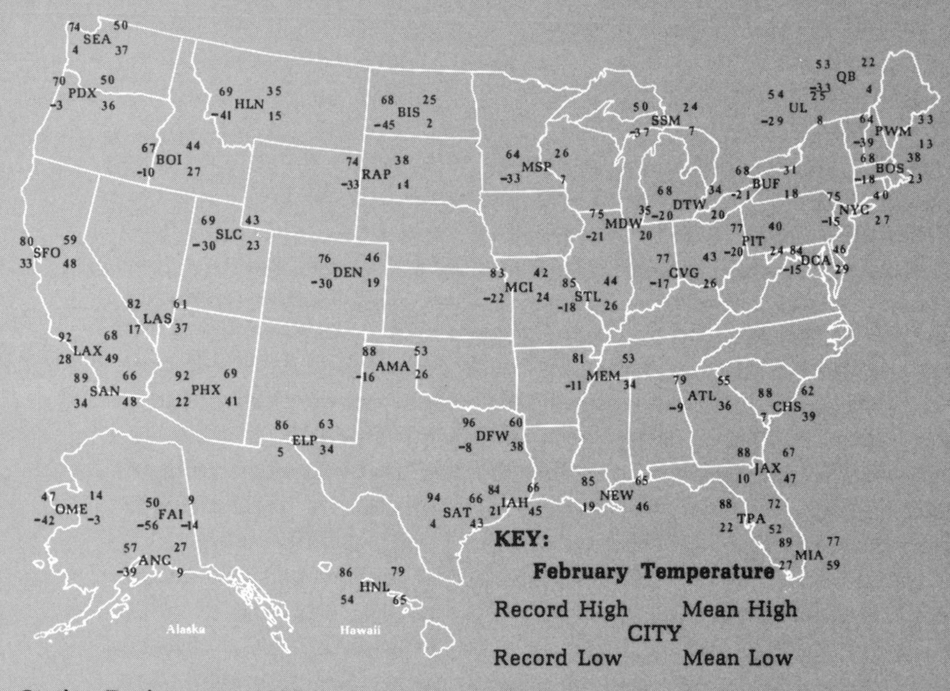

Station Designators: **AMA** Amarillo TX; **ANC** Anchorage AK; **ATL** Atlanta GA; **BIS** Bismarck ND; **BOI** Boise ID; **BOS** Boston MA; **BUF** Buffalo NY; **CHS** Charleston SC; **CVG** Cincinnati OH; **DCA** Washington DC; **DEN** Denver CO; **DFW** Dallas-Fort Worth TX; **DTW** Detroit MI; **ELP** El Paso TX; **FAI** Fairbanks AK; **HLN** Helena MT; **HNL** Honolulu HI; **IAH** Houston TX; **JAX** Jacksonville FL; **LAS** Las Vegas NV; **LAX** Los Angeles CA; **MCI** Kansas City MO; **MDW** Chicago IL; **MEM** Memphis TN; **MIA** Miami FL; **MSP** Minneapolis-St. Paul MN; **NEW** New Orleans LA; **NYC** New York NY; **OME** Nome AK; **PDX** Portland OR; **PHX** Phoenix AZ; **PIT** Pittsburgh PA; **PWN** Portland ME; **QB** Quebec QUE; **RAP** Rapid City SD; **SAN** San Diego CA; **SAT** San Antonio TX; **SEA** Seattle WA; **SFO** San Francisco CA; **SLC** Salt Lake City UT; **SSM** Sault Ste. Marie MI; **STL** St. Louis MO; **TPA** Tampa FL; **UL** Montreal QUE; **WG** Winnipeg MAN; **YC** Calgary ALB.

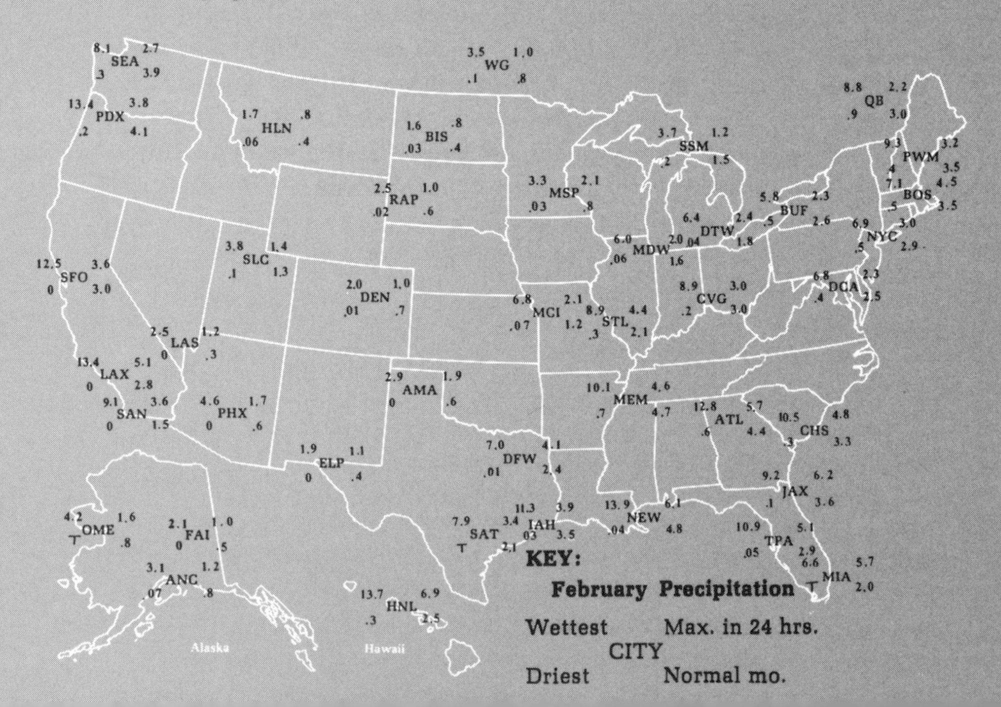

2 1951 Severe storm spread thick icy sheath from Texas to Pennsylvania; Tennessee suffered most, with communications and utilities interrupted for 7 to 10 days; $100 million loss.

3 1844 Channel sawed through ice of Boston Harbor released Cunarder S.S. *Britannia*, which sailed for England with 30,000 delayed letters.

1917 Downtown Miami's lowest temperature in Weather Bureau history (since 1911): 27°F (−3°C).

4 1961 Great Northeast Snowstorm, the third of the season, spread deep mantle; up to 40 in (102 cm) fell at Cortland, N.Y.

5 1887 San Francisco's greatest snowstorm: 3.7 in (9.4 cm) downtown, 7 in (18 cm) on western hills of city.

6 1807 Famous Cold Friday in Midwest and South: temperature did not rise above 0°F (−18°C) in Ohio and Kentucky.

1978 Greatest snowstorm in modern history of southeast New England; record 27.2 in (69 cm) at Boston; 38 in (97 cm) in Rhode Island; Boston snowbound, all travel banned for a week by governor's decree.

7 1861 Spectacular temperature wave: At Gouverneur, N.Y., thermometer fell from 30°F (−1°C) on February 7 to −40°F (−40°C) the next day, rose to 55°F (13°C) on February 10; Hanover, N.H., had afternoon 40°F (4°C) on February 7 and morning −32°F (−36°C) on the following day.

8 1835 Great Arctic Outbreak into the Southeast brought coldest ever experienced; 0°F (−18°C) at Charleston and Savannah; 8°F (−13°C) near Jacksonville, Fla.; nascent citrus industry ruined.

9 1870 President Grant signed measure creating a federal meteorological service, which was assigned to the U.S. Army Signal Service to organize and operate; observations began on November 1, 1870.

10 1973 Big snowstorm in Southeast: 24 in (61 cm) in South Carolina, 18 in (46 cm) in Georgia; north-south Interstate blocked several days.

11 1899 Greatest of all arctic outbreaks commenced with temperature dropping to −61°F (−52°C) in Montana.

12 1899 Coldest morning of modern record, eastern Plains and Texas: Kansas City −22°F (−30°C), Fort Worth −8°F (−22°C). In East, Washington hit −15°F (−26°C). Downtown Charleston, S.C., had its greatest modern snowfall of 3.9 in (10 cm).

13 1784 Ice floes blocked Mississippi River at New Orleans, then passed into Gulf of Mexico; only similar occasion was in 1899.

1899 Coldest morning ever along Gulf Coast: 6.8°F (−14°C) at New Orleans, −1°F (−18°C) at Mobile, 7°F (−14°C) at Pensacola, −2° (−19°C) at Tallahassee, Florida's all-time low.

14 1899 Great Eastern Blizzard of '99; Washington had 20.5 in (52 cm), to reach snow depth of 34 in (86 cm); Cape May, N.J., measured storm total of 34 in (86 cm), making 41 in (104 mm) on ground.
1940 St. Valentine's Day blizzard in southern New England; 10–17-in (25–40-cm) snowfall and whole gales; many stranded in downtown Boston after Sonja Henie ice show.

15 1895 Heaviest snowstorm ever in west Gulf region; 24 in (61 cm) at Rayne, La., 20 in (51 cm) at Houston, Tex., 6 in (15 cm) at Brownsville, Tex.

16 1943 Spectacular cold wave set Connecticut record, −32°F (−36°C) at Falls Village; other low readings were −30°F (−34°C) in Massachusetts and −46°F (−43°C) in Vermont.

17 1748 Coldest day in colonial history of South: 10°F (−12°C) at Charleston, S.C.
1958 Great snowstorm of midtwentieth century in Northeast: 36 in (91 cm) in Poconos; 30 in (76 cm) in Catskills; 30 in (76 cm) in interior of New England; Boston's most intense snowfall to that time, 19.4 in (49 cm) in 24 hours at airport.

18 1899 San Francisco's warmest ever in February: 80°F (27°C).

18 1979 Great Snowstorm of '79 at Washington, D.C.; 18.7 in (47 cm) in 18.5 hours at airport; total depth 23 in (58 cm), both third greatest; Baltimore airport had 20 in (51 cm).

18 1980 Series of six major storm fronts swept California from 13th to 21st, hitting the southern part the hardest where 30 deaths resulted from floods, mudslides, and traffic accidents; L.A. Civic Center 12.75 in (324 mm), and Mt. Wilson (805 mm).

19 1884 Massive outbreak of 60 tornadoes in Southeastern states killed estimated 800 persons; Georgia and Carolinas suffered most.
1888 Tornado at Mount Vernon, Ill., killed 16; path 62 miles (100 km) long.

20 1898 Eastern Wisconsin's biggest snowstorm: Racine 30 in (76 cm) and drifts of 15 ft (4.6 m) at Milwaukee.

21 1918 Spectacular chinook wind at Granville, N.D.; temperature spurted from morning −33°F (−36°C) to afternoon 50°F (10°C), a rise of 83°F (46°C).
1971 Four tornadoes in Louisiana and Mississippi killed 121, injured 1600, did $19 million damage; combined total of 380 miles in ground contact over five-hour period.

22 1773 "Memorable Cold Sabbath" in New England; many froze extremities while going to church.

23 1802 Great snowstorm raged on New England coast; depths of 48 in (122 cm) north of Boston; three large Indiamen out of Salem wrecked on Cape Cod.

24 1852 Susquehanna River ice bridge at Havre de Grace, Md., commenced to break up after 40 days' use; 1378 loaded freight cars hauled across on rails laid on ice.

25 1921 Downtown Los Angeles reached 91.6°F (33°C), warmest ever in February.

26 1972 Heavy rains caused slag-dam failure on Buffalo Creek in West Virginia; settlements in narrow hollow wiped out; 118 killed.

27 1717 First of series of four snowstorms in 10 days making up "The Great Snow of 1717"; finally left 36 in (91 cm) on ground at Boston; city snowbound for two weeks; much more fell to north.

28 1900 Massive snowstorm set records from Kansas to New York State; 17.5 in (44 cm) at Springfield, Ill.; 43 in (109 cm) at Rochester, N.Y.; 60 in (152 cm) in Adirondacks; 31 in (79 cm) at Northfield, Vt.

29 1748 End of heaviest snowstorm of "The Winter of the Deep Snow" in New England left 30 in (76 cm) on ground at coastal Salem.

Snow

Announced by all the trumpets of the sky,
Arrives the snow; and, driving o'er the fields,
Seems nowhere to slight; the whited air
Hides hills and woods, the river and the heaven,
And veils the farm-house at the garden's end.

—Ralph Waldo Emerson, "The Snowstorm"

Oh! the snow, the beautiful snow
Filling the sky and the earth below:...
Beautiful snow, from the heavens above,
Pure as an angel and fickle as love!

—John Whitaker Watson, "Beautiful Snow"

A snowfall consists of myriads and myriads of minute ice crystals whose birthplace lies in the subfreezing strata of the middle and upper atmosphere when there is an adequate supply of moisture present. At the core of every ice crystal is a minuscule nucleus, a

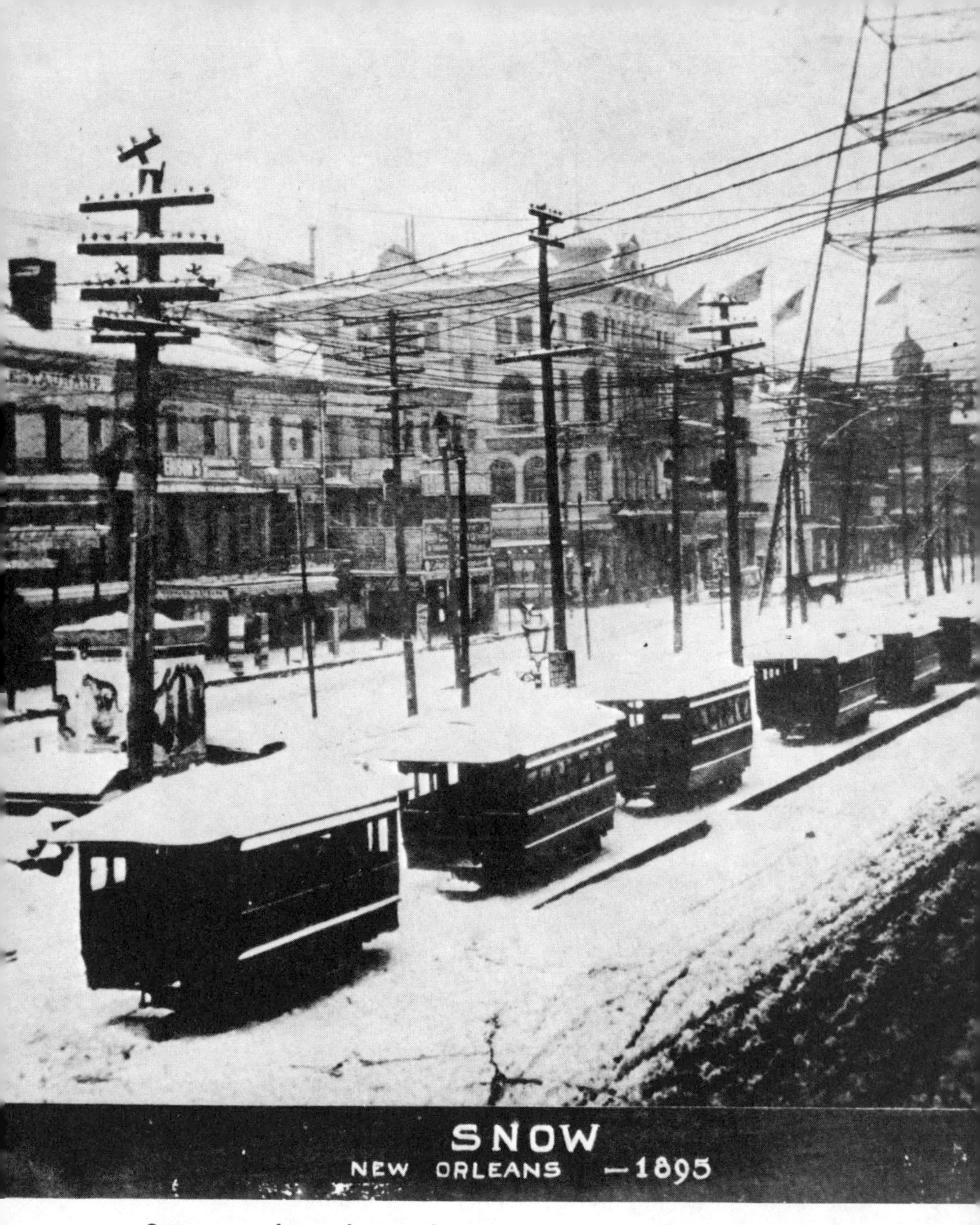

Snowstorm dropped 8.2 inches on New Orleans, February 14–15, 1895, stalling horse cars at corner of Crawford and Hill streets. Louisiana Historical Society photo.

solid particle of matter of varying composition, around which moisture condenses and freezes. Liquid water droplets floating in the supercooled atmosphere and free ice crystals cannot coexist within the same cloud since the vapor pressure of ice is less than that of water. This enables the ice crystals to rob the liquid droplets of their moisture and grow continuously by accretion. The process can be very rapid, quickly creating sizable ice crystals, some of which adhere to each other to create an aggregate of ice crystals or a snowflake. Simple flakes possess a variety of beautiful forms, usually hexagonal, though the symmetrical shapes reproduced in most photomicrographs of snowflakes are not found in actual snowfalls. Snowflakes consisting of broken fragments and clusters of adhering ice crystals are the forms making up a typical snowstorm.

For a snowfall to continue once it starts, there must be a constant inflow of moisture to feed the hungry condensation nuclei. This is supplied by the passage of an airstream over a water surface and its subsequent lifting to higher regions of the atmosphere. The Pacific Ocean is the source of moisture for most snowfalls west of the Rocky Mountains, while the Gulf of Mexico and the Atlantic Ocean feed water vapor into the air currents over the central and eastern sections of the country. The thermodynamics and hygrodynamics of large cyclonic storm systems govern snow making through the successive processes of warming, convection, cooling, condensation, and precipitation. Areas to the lee of the Great Lakes experience their own unique lake-effect storms, employing a variation of the above on a local scale, and mountainous sections or rising terrain can initiate snowfalls by the orographic lifting of a moist airstream to initiate precipitation.

A single minute snow crystal is seemingly incapable of being an element of force or destruction, but when combined in countless quantities they make a profound impact on many aspects of human life. When floating in the upper atmosphere, snow crystals cause beautiful halos and other optical displays surrounding the sun and moon. When present in the middle layers of the atmosphere in cumulonimbus clouds, they contribute to the creation of electrical charges, resulting in lightning, thunder, and static phenomena. When in the clouds high over the equatorial zones, they melt into raindrops in their descent and cause the heavy downpours typical of low latitudes. When in the temperate zones in winter, they fall to earth in an unmelted state, resulting in heavy snowfalls that transform the landscape and disrupt transportation and normal economic life. When falling in arctic regions, they contribute an essential ingredient to the dreaded blizzard of the frozen wastes of the polar sea and the continental tundra.

While accumulating on mountain slopes, snowflakes compact and create the snowpack so vital to the nation for summer water supply. Occasionally, the descent of a mass of snow institutes an irresistible avalanche, leaving death and destruction in its wake. On the more positive aspect, the coming of the snow season is welcomed by youth and the more hardy adults by creating the scene for the enjoyment of many winter sports. As the snows accumulate on the ground, so do the contributions that the winter wonderland annually makes to the national economy mount.

Unusual Snowstorms

Up in Vermont when a summer visitor asks a native what he does in winter when it snows, he usually answers laconically with his Yankee twang, "We just let it snow." Anyone along the northern border of our country, from Madawaska, Maine, to Blaine, Washington, expects it to snow frequently and to snow hard during the winter season.

Residents of more southerly climates, however, have been conditioned differently. Often they suffer from *nivia tremens* when a few white flakes filter down from the sky. So let's consult the "oldest inhabitant," that perennial source of weather wisdom, and explore the situation below the Mason-Dixon Line, where snows are somewhat of a rarity, and also along the Pacific Coast, where snowfalls along with everything else are always "unusual."

Lower Texas Coast — Snowflakes have fallen along the Coastal Bend country of Texas in measurable amounts on only three occasions since the Civil War, and all were prior to 1900. Brownsville, close to the mouth of the Rio Grande at 25°54′N, had its memorable "Big Snow" on February 14, 1895, when the official measurement was 3.0 in (8 cm) but the local newspaper mentioned 6.0 in (15 cm). Nothing like this had occurred since January 1866, according to "old Chicanos." After 1900 there have been at least ten occasions when traces of snow have been reported at Brownsville, but no measurable amounts accumulated on the ground. Moving farther north along the coast, Corpus Christi (27°46′N) had two substantial snowstorms in the olden days: in 1895, 4.3 in (11 cm) and in 1897, 5 in (13 cm). Those who remembered the Alamo could recall nothing like these, and their descendants in this century have not seen the likes. Recently, small measurable amounts have fallen on two occasions: 0.2 in (0.5 cm) in January 1973 and 1.1 in (2.8 cm) in February 1973.

Upper Texas Coast — The most memorable of snowstorms along the upper Texas coast comprises what is probably the greatest deep-snow anomaly in the recorded climate history of the United States. On February 14–15, 1895, a small coastal low-pressure area dropped a wide canopy of snow that centered in the northwest corner of the Gulf of Mexico. Coastal Galveston reported an unprecedented 15.4-in (39-cm) snowfall, and a short distance inland at Houston the official measurement went into the books as 20 in (51 cm), though the press at the time stated that the total fall was 22 in (56 cm). Descendants of the heroes of San Jacinto could remember nothing like this having occurred.

Nothing exciting in the snowfall line occurred again until February 1960, when Houston had a fall of 4.4 in (1.2 cm), and unofficial amounts of 8 to 12 in (20 to 31 cm) were reported in the Orange–Beaumont area of the extreme northeast coastal strip. The year of 1973 was distinguished for three measurable snows at Houston, totaling 2.0 in (5 cm) in January and 2.8 in (7 cm) in February — trifling amounts compared with the "Big Snow of '95."

Central Gulf Coast — The heavy snow pattern of February 1895 reached its maximum in western Louisiana where Rayne Parish reported 24 in (61 cm) and Lake Charles 22 in (56 cm). Even the "oldest Cajun" could not remember his grandpère telling of anything like this, and nothing like it has happened since. Snow fell on this occasion — February 14–15, 1895 — even at the very mouth of the Mississippi River. Recent years have brought two snowfalls to the Louisiana coastal area: Lake Charles measured 5.0 in (13 cm) in February 1960 and 4.0 in (10 cm) in January 1973.

The New Orleans snow story shows that measurable amounts have fallen in 1784, 1812, 1834, 1852, 1867, 1881, 1895, 1899, 1958, 1963, and 1973. The greatest amount, 8.2 in (21 cm), came in the big storm of February 14–15, 1895. In the "Sugar Bowl" snowstorm on December 31, 1963, 4.5 in (11 cm) of snow and sleet greatly complicated the task of driving to the event. "Old Creoles" had seen nothing like this. The winter of 1972–73 had the distinction of bringing two measurable snows to the Crescent City, though both were of trifling amounts.

Mobile's greatest snowstorm dates way back to January 1881, when 5.0 in (13 cm) fell. In recent years the Alabama port has seen 3.6 in (9 cm) on February 9, 1973, and a total of 1.9 in (5 cm) in two storms in January 1977.

Pensacola's maximum snowstorm seems to have been 3 in (8 cm), falling in the general storm of February 1895. Snow has been experienced at the home of naval aviation in 28 of the 84 years of record, but there have been only eight occasions with measurable amounts. The most recent were 1.9 in (5 cm) in February 1973 and 2.5 in (6 cm) in January 1977.

Florida Peninsula — Snowflakes have been seen in the air southward as far as the mainland of the Florida peninsula extends. This was not true until the "Great Snow of '77" brought a trifling trace of snow on January 19, 1977, to the Miami area and as far south as Homestead, on the latitude where the Florida Keys commence. The "oldest Seminole" in the Everglades was consulted and could not recall anything like this in his folklore.

On the west coast the farthest penetration of the Snow Kingdom in official records has been in the Punta Rassa–Sanibel Island area at 26°30′N, where a trace of snow was reported by the local observer on December 1, 1876. Nearby Fort Myers has seen flakes in the air on only four occasions in one hundred years of records. No snow has ever been reported at Key West, the southernmost extremity of the lower 48 states, where the lowest official temperature has been 41°F (5°C).

Snow covers Florida palms near St. Augustine, February 2–3, 1951. U.S. Weather Bureau photo.

Tampa, on the Gulf Coast at 27°48′N, has seen the ground whitened on three occasions when there was enough to scrape together to make snowballs: in January 1886, February 1899, and January 1977. On the most recent occasion the measurement was 0.2 inch (0.5 cm), but Lakeland, not far to the east, received a full inch (2.5 cm).

For extent and depth, the greatest snowstorm in Florida's history occurred on February 12–13, 1899, during the severest arctic outbreak of modern times into the South. Measurable snow fell on this occasion as far south as Tampa on the west coast, Lakeland–Orlando in the center of the peninsula, and Daytona Beach on the east coast.

Florida's second greatest snowstorm occurred on February 12–13, 1958, when as much as 3.0 in (8 cm) was reported at several stations in the north and measurable amounts fell generally down to latitude 30°N. The next measurable snow came on February 9–10, 1973, with 3.3 in (8 cm) at Milton in the northwest corner of the state, with unofficial reports of 6 to 8 in (15 to 20 cm) in Santa Rosa County, along the Alabama border.

Florida's fourth outstanding snowfall was the celebrated occasion on January 18–19, 1977, when snowflakes were observed well below Miami. In a shower-type precipitation at night and in the early morning, flakes were observed officially at the weather stations at Homestead, Miami Beach, and Palm Beach, for the first time in history at each location. Windshields at Fort Lauderdale were photographed with enough snow to require wiping. In the central section of the state Saint Leo and Lakeland reported one inch (2.5 cm).

Jacksonville, near the northeast corner of the state, with a record dating back almost a century and a half, has had only two occasions with snow that required a ruler measurement: 1.9 in (5 cm) on February 13, 1899, and 1.5 in (4 cm) on February 13, 1958. Recently, traces of flakes in the air were observed in February 1973, March 1975, and January and February 1977, but they did not whiten the ground.

In the panhandle area of Florida, along the northern border, where continental conditions occasionally prevail, snowstorms have been reported in 1773, 1800, 1852, 1886, 1887, 1892, 1895, 1899, 1943, 1954, 1955, 1958, 1973, and 1977. In the stretch from Pensacola over to Tallahassee, storms in the range of 3 to 4 in (8 to 10 cm) have been recorded in the past 130 years, but none exceeded the 5.0-in (13-cm) snowfall reported on January 11, 1800, by surveyor Andrew Ellicott when surveying the southern boundary of the United States between Georgia and Spanish Florida.

Atlantic Southeast Coast, Savannah — The seaport of Savannah has had only four snowstorms in the past hundred years that required a precise measurement: on January 18, 1893, a full one inch (2.5 cm) accumulated; on February 12–13, 1899, the depth was 2.0 in (5.1 cm); on February 8, 1968, the airport measured 3.6 in (9.1

cm); and on February 10, 1973, the amount was 3.2 in (8.1 cm). A recent sneak storm gave downtown Charleston 4.0 in (10 cm) and the Airport 3.8 in (10 cm) on December 26–27, 1980. Only the coastal strip had measurable snowfall. Savannah reported only a trace. But in the old-fashioned days it was different. In January 1800, the *Columbian Museum and Savannah Advertiser* carried the following:

> A snow storm, accompanied by high wind, commenced in this city last Thursday evening, the 9th, and continued all that night, Friday and Friday night incessantly, until Saturday evening, which left the snow above 18 inches deep, and is not half dissolved yet [Tuesday, January 14] although it has been clear sunshining weather ever since.

Atlantic Southeast Coast, Charleston — Looking back in the history books of early Carolina reveals that it was much snowier at Charleston in the olden days than it has been recently. Substantial snowfalls were mentioned in the years 1743, 1748, 1773, 1790, 1792, 1800, and 1803. There were four storms in the winter of 1799–1800, when Charleston had its all-time snowiest month, with a total of 14.0 in (36 cm). After this and the 1803 event, no major snows fell at Charleston for almost a hundred years, until the big Southern Snowstorm on February 12, 1899, dropped 3.9 in (10 cm). Only centenarians in the Palmetto City could remember anything like that. In the present century there have been measurable snows of lesser depth in 1912, 1914, 1915, 1917, 1943, 1966, 1968, 1969, 1973, 1976, 1977, and 1980. The increase in measurable snow events in recent years represents the record at the airport, which is located 10 mi (16 km) inland from downtown Charleston and experiences cooler temperatures. The biggest snow there came on February 9–10, 1973, with 7.1 in (18 cm), and amounts were double that farther inland.

Los Angeles Metropolitan Area — In downtown Los Angeles at the Civic Center measurable snow has occurred only twice in modern times. The "Big Snow" came on January 15, 1932, when 2.0 in (5 cm) were judged to have fallen at the city weather station, though the measured depth on the ground never exceeded one inch (2.5 cm). Even the beaches at Santa Monica had a blanket of white on this occasion. Again on January 11, 1949, a white mantle descended on the coastal plain, with the official measurement placed at 0.3 in (0.76 cm). Much greater depths covered the hills, and some mountain passes were closed until plows arrived. On twenty-two other occasions in the past hundred years snowflakes have been observed in downtown Los Angeles, but there were no measurable accumulations.

San Francisco Bay Area — San Francisco's big snow event came in 1887 on February 4–5, when a rain turned to wet snow and covered downtown areas with 3.7 in (9 cm) officially, and the hills in the western sections were whitened with as much as 7.0 in (18 cm). Great snow battles broke out across the city, with cable car passengers and Chinese residents being the principal targets; a number of arrests were made for unseemly conduct. San Francisco's No. 2 snowstorm had come earlier in the same decade, when 3.5 in (9 cm) fell during the afternoon of New Year's Eve, December 31, 1882.

In this century there have been only two occasions with measurable snowfall: on December 11, 1932, a recording of 0.8 in (2 cm) was made downtown, and on January 21, 1962, the airport south of the city at sea level was covered to a depth of 1.5 in (4 cm). The hills surrounding the Bay, of course, are often whitened, but the snow line usually stays above 500 ft (152 m). In 1976 the higher ground was covered by flakes on three occasions. The "oldest Argonaut" could not remember seeing such frequent snows.

Big Snows in Big Cities

Chicago's Big Snow in January 1967 — Never in the history of the Midwest has such a mass of snow fallen on so many people in such a short time as descended from the skies over Chicagoland on January 26–27, 1967.

A snowfall of 20 in (51 cm) or more in a single storm places a locality in the elite class of American urban snowfall statistics. During the first eighty-two years of its official records, Chicago never produced a qualifying total. Only two storms had exceeded

15 in (38 cm) in that long period. Thus the admission of the second largest city in the country to the exclusive "20 Inches Snowstorm Club" was a historic event, indeed. Further, the unprecedented paralysis of the entire urban complex and the traffic blockade of surface and air transportation at the crossroads of the Midwest caused unexpected economic distress and unanticipated disruption of normal life. These factors raised the 1967 storm at Chicago to the "great" category and a deserved place among famous American snowstorms.

The previous record for a single snowfall at Chicago was held by the post-equinox storm on March 25–26, 1930, when 19.2 in (49 cm) fell in a forty-eight-hour period. National Weather Service records of snowfall at Chicago date from the season of 1884–85. A search of previous records of the Signal Corps and of individuals reveals no outstanding snowstorm of the past, nor do the files of the *Chicago Tribune* (the "World's Greatest Newspaper") make any reference to the time-honored reputation of a historic storm. Thus, it is safe to assume that the 1967 snowstorm was the greatest in all Chicago history.

Chicagoans awoke on Thursday morning, January 26, 1967, to a snowy, blowy day. Flakes had begun to filter down about 5:30 A.M., and already there were accumulations. By midmorning 4 to 5 in (10 to 13 cm) lay on the ground, and it was snowing heavily. Northeast gales accompanied the flakes, driving them along at a rate that mounted to a peak of 53 mi/h (85 km/h). Temperatures ranged from 26 to 29°F (−3 to −2°C) during the entire storm period, so it did not rate as a blizzard. By the evening rush hour there were about 12 in (31 cm) on the ground, and the heavy fall continued. When it finally stopped at 10:10 A.M. next morning, after 29 hours, the official measurement was 23 in (58 cm), which constituted a record for any single storm at Chicago, and the 19.8 in (50 cm) falling in 24 hours set a record in that category. Along with additional snows in early February, the snow mounted to a record depth of 28 in (71 cm) on February 7.

The total cost of removing snow in the winter of 1966–67 was over $5 million, compared with a normal seasonal outlay of $500,000. The total cost of business losses during the storm was estimated at $150 million, which represented the amount of goods

2232

and services that went unsold. The biggest hindrance in clearing the streets was the effect of abandoned cars, whose removal caused the cleanup to take ten times longer than normal.

Though such a mighty snowstorm as January 1967 brought is supposed to occur once in a hundred years, just a dozen years later came along a worthy rival. On January 12–14, 1979, Midway Airport measured a fall of 20.7 in (53 cm), which contributed to an all-time record depth on the ground of 29 in (74 cm). The *Chicago Tribune* estimated that the economic losses throughout the area for the first week following the storm amounted to $1.25 billion.

New York's Big Snow — Most residents of the city were sleeping off an active Christmas Day when, unbeknownst to them, a light snow began to fall at 3:20 A.M. on the morning of December 26, 1947. The forecasts appearing in the papers and broadcast by radio on Christmas morning had made no mention of snow for the morrow, but in midafternoon the Weather Bureau issued a bulletin calling attention to current indications for "increasing cloudiness, some snow likely after midnight . . . Friday [the twenty-sixth] light snow ending in forenoon."

For the twenty-four hours after the snow started to fall, the forecasters had a hard time keeping up with the amount of snow rapidly accumulating on the ground, and almost hour by hour upgraded the amount of snow expected. At 7:00 A.M. the depth amounted to 1.8 in (5 cm) — by 12:00 noon to 11 in (28 cm) — by 6:00 P.M. to 22.5 in (57 cm) — by midnight to 25.6 in (68 cm) — and when it ended at 3:05 A.M. to 25.8 in (66 cm) at Battery Park at the southern end of Manhattan Island. The official count at Central Park came to 26.4 in (67 cm), and this is the total now assigned to the "Big Snow."

The snow-making disturbance had formed over Alabama early on the twenty-fourth in the southern end of an eastward-moving trough. When snow began to fall over New York City on the early morning of the twenty-sixth, the low center was over the

Opposite: Cars stranded and abandoned in blizzard, South Lake Shore Drive between Twelfth and Twenty-fifth streets, Chicago, January 27, 1967. Chicago Tribune photo.

ocean east of the Maryland-Delaware border and was heading almost directly north. During the daylight hours the trough from the west and the storm center from the south merged to form a large disturbance. At the same time a developing anticyclone over Quebec served to block the forward progress of the center, causing it to deepen. When the center passed over the extreme eastern portion of Long Island about 7:30 P.M., the barometric pressure dropped to 29.24 in (99.0 kPa).

The intensification of the storm center kept the precipitation going much longer than normal. Ordinarily a coastal storm will produce heavy precipitation for about six hours, but the Big Snow kept going at full rate for eighteen hours. Throughout the daylight hours of December 26 snow fell at the rate of 1.8 in (4.6 cm) per hour, and at times in late afternoon at an hourly rate just under 3.0 in (7.6 cm).

The impact of the deep snow on traffic was staggering. Streets and highways were rendered impassable to vehicles; at least 10,000 automobiles were abandoned in the snow; and bus, street car, elevated, and subway lines were almost totally paralyzed. All airplane flights were suspended, harbor shipping was brought to a halt, and railroad service was canceled or subjected to long delays. New York City appropriated $7 million in addition to the already allotted $1.3 million for snow removal. The Department of Sanitation hired 24,000 extra workers to serve as snow gangs, and threw 3200 pieces of equipment into the struggle against the huge snowbanks. Economic losses during Christmas Week were incalculable.

Snowbound in Boston: February 1978 — When thinking of wintertime in New England, one conjures pictures of "old-fashioned" snowstorms and the countryside buried deep in snow. Yet within the hundred years of official record keeping, the principal city of the region had never experienced a concentrated heavy snowstorm, defined as 20 in (51 cm) or more falling within a 48-hour period. Other New England cities have been buried in such intense snowfalls, and even some cities in the Deep South have experienced the like. It remained for the winter of 1977–78 to produce not one, but two falls of 20 in or more at Boston within a period of 18 days in late January and early February 1978.

Boston was struck by a true blizzard on January 20, 1978, in an intense storm that brought winds gusting up to 62 mi/h (100 km/h) and dumped 21 in (53 cm) on Logan Airport within 24 hours. Suburban districts reported up to 24 in (61 cm). Many buildings throughout the state suffered damage from collapsing roofs. This was followed within a week by a high windstorm without snow that toppled trees and signs throughout the state.

After the atmosphere settled down from these two convulsions, eastern New England lay under the control of a persistent low-pressure system over the Atlantic Provinces of Canada and a high-pressure area over the Great Lakes and Hudson Bay. This assured a steady flow of bitterly cold air from the central Arctic region, keeping temperatures low without additional precipitation. Late on Sunday, February 5, a low-pressure trough moving eastward from the upper Mississippi Valley translated its energy to a new cyclonic center off the Virginia Capes, and with rapidly developing energy the new storm moved north-northeast on the sixth to a position south of Long Island. Here it came under the blocking influence of the high pressure extending across central and eastern Canada, where the barometer stood at the elevated figure of 31.07 in (105.2 kPa) over northwest Ontario.

Snow began to fall in New York City early on Monday morning, February 6, and soon spread northeast through southern New England. Some flakes were reported at Boston after 7:00 A.M., but they did not become steady until 10:20, and even then the fall was very light. The Boston forecasters saw the potential for a major storm; they expected the center to intensify quickly and move east-northeast, passing close to Nantucket and Cape Cod, a track that makes for big snows in the Boston area. Predictions of 8 to 16 in (20 to 41 cm) were issued on Monday morning.

The storm began to make itself felt in the Boston area soon after noon, with the wind picking up to a steady 35 mi/h (57 km/h) and gusting above 40 mi/h (65 km/h). Yet the light snow falling during the noon hour continued through the afternoon with only a very small accumulation. At 5:30 P.M., however, the fall became heavy and the real storm was on. At 7:00 P.M. 3 in (8 cm) of new

WINTER

snow were measured, and by midnight a new accumulation of 8 in (20 cm) lay on top of the original morning base of 2 in (5 cm). The peak wind gust of the entire storm came about 9:00 P.M. at 79 mi/h (128 km/h) out of the northeast.

The heaviest snow of the storm commenced soon after midnight and fell at the rate of about one inch per hour until 7:30 A.M. on Tuesday morning, when 16 in (41 cm) were reported. Snowfall turned light during most of the morning before becoming heavy again between 1:00 and 4:00 P.M. and again between 4:30 and 5:00 P.M. Then it slacked off to light, preparatory to ending completely at 11:05 P.M. By Tuesday noon, 22 in (56 cm) of new snow had fallen, and by 7:00 P.M. the total figure was raised to 27 in (69 cm). No sleet was reported at Logan Airport, as often occurs in long-duration snowstorms along the coast.

The snowfall had continued for a total of 36 hours and 45 minutes. During this time three all-time records were set for the city of Boston: The greatest 24 hours of snow, 23.6 in (60 cm), were registered following 7:00 P.M. on Monday. The greatest single storm total of 27.5 in (70 cm) fell on top of a 2-in (5-cm) base, and this gave a total depth of 29 in (74 cm) at the conclusion.

A snow emergency was declared within Boston at 4:00 P.M. on Monday in anticipation of a large fall. Logan Airport was closed late in the evening, when wind and blowing snow made the runways unsafe, and a state of emergency was declared for Massachusetts on Tuesday morning by Governor Michael Dukakis. Boston became snowbound early Tuesday and remained so for five days. All but emergency vehicles were banned from the streets. Schools were closed. Practically all businesses shut down, either from lack of customers or failure of supplies to arrive. Social and sporting events were canceled. All transients were urged to leave town, and none was allowed into the area. It was Boston's most memorable snow experience since the legendary "Great Snow of 1717."

Opposite: Winter on Winter Street, Bangor, Maine. Bangor Daily News photo.

K.P.R.

March

The stormy March has come at last,
With wind, and cloud, and changing skies;
I hear the rushing of the blast,
That through the snowy valley flies.
— William Cullen Bryant, "March"

The first day of spring is one thing, and the first spring day is another. The difference between them is sometimes as great as a month.
— Henry Van Dyke, Fisherman's Luck

Opposite: Sailing car on the Kansas Pacific Railroad. Kansas Pacific Railroad. Kansas Historical Society photo.

Lion-like March comes in, hoarse, with tempestuous breath." So William Dean Howells described the blustery month that is aptly named after the Roman god of war. The forces of the North are in frequent combat with the forces of the South for possession of the great American heartland during this transition month from winter to spring. All sorts of meteorological warfare and frolics break out when warm, moist, tropical air from the Gulf of Mexico and cold, dry, polar air from central Canada come into juxtaposition. Balmy temperatures well above 68°F (20°C) may prevail on the eastern side of a weather front, while freezing conditions with a wind chill near 0°F (−18°C) are occurring on the western side. Spring tornadoes spin northward on one side, while late-winter blizzards roar south on the other. With vastly increased supplies of heat and moisture, the March weather elements marshal greatly contrasting airstreams and produce some of the most spectacular weather of the year.

The principal event on the March calendar is the vernal equinox, arriving on or about the twenty-first, the date marking the commencement of astronomical spring, though meteorological spring may not yet have arrived in your locality. The term *equinox* is derived from the Latin *aequinoctium,* or equal nights, i.e., when day and night are of the same duration. On that date in March the sun, advancing northward, crosses the celestial equator, and its direct overhead rays first fall on the Northern Hemisphere across northern Brazil, southern Columbia, and Ecuador. As Henry Van Dyke pointed out, the event has little effect in determining the beginning of meteorological spring but merely confirms that the solar angle is increasing, making more heat energy available for the atmospheric engine to produce a more varied weather fare.

In its most southerly position, the Pacific High continues as a strong weather map feature during March, but the subcenter area over Idaho and Utah weakens in strength and declines as a blocking influence on storm movement. The Aleutian Low is less vigorous and is farther west than in February; its center is found over the extreme western Aleutian chain with a mean central pressure of approximately 29.69 in (100.5 kPa), a diminution in depth of 0.15 in (0.5 kPa) from the winter's low.

In the Atlantic Ocean, the Azores–Bermuda High remains off the West African coast with a center southwest of the Azores near latitude 31°N and an extension west toward the Georgia coast. The center of the Iceland–Greenland Low lies off the southeast tip of Greenland; its central pressure has risen 0.30 in (1.0 kPa) from the winter's minimum, an indication of the decline in the violence and frequency of severe storms that spring brings.

The principal jet stream in March enters the country over Oregon and dips southeast to Missouri before heading east, close to the February track toward the Virginia coast. The low-latitude jet entering over the west coast of Mexico has disappeared as a major factor; as a result, the frequency of storms in Southern California diminishes.

March is a transition month, when the storm tracks retain many of the winter aspects of January and February yet at the same time begin the change to conditions that will become firmly established in April and May. Storms form with increasing frequency over such inland areas as the Rio Grande Valley and along the lee slopes of the Appalachian Mountains. Storm systems over the eastern United States and the North Atlantic area are often influenced by the blocking effect of high-pressure areas over eastern Canada, which tend to remain stationary for many hours and days, preventing coastal storms from moving northeast on a normal route and resulting in prolonged periods of storminess in the Middle Atlantic and New England states. The storm track from the Gulf of Alaska over the western mountains to northern Alberta remains the primary continental storm path in the West, and the more southerly track from the central Pacific Ocean into Washington and Oregon is still active.

Although the principal anticyclone tracks during March retain most of the characteristics of the winter months, several features typical of spring are introduced. The offshoots of the Pacific High move eastward at more northerly latitudes than in winter and often pursue a new route through the Ohio Valley to the central New England coast. In Canada an anticyclonic area centers over Saskatchewan, and farther east a path from west of Hudson Bay leading directly into northeastern United States becomes an increasingly important factor in influencing spring weather. The in-

Locomotive and first cars of freight train stalled in a cut while track in rear was swept clear during great blizzard of March 3–5, 1966, near Bismarck, North Dakota. NOAA photo.

creased elevation of the sun and longer hours of sunshine make a definite impression across all the lower 48 states. The winter's cold pole in North Dakota shows a warming in March of 12.4F (6.9C) degrees over February. At the northern corners of the country the varying effects of wind direction are evident: At Eastport, Maine, where continental influences prevail at this season, the March increase over February amounts to 10.6F (5.9C) degrees, while at Seattle, Washington, under maritime controls, the rise is only 1.4F (0.8C) degrees. The mean isotherm of 60°F (16°C), formerly confined to the southern half of Florida and southeast Texas, encompasses all the Gulf Coast; the 32°F (0°C) mean line retreats to the northern Plains, northern New England, and the higher mountain elevations of the West.

Temperature extremes in March: maximum 108°F (42°C) at Rio Grande City, Texas, on March 31, 1954; minimum −50°F (−46°C) at Snake River, Wyoming, on March 17, 1906.

March is a wet month in the eastern half of the country, and many mighty floods have occurred. The 4.0 in (102 mm) mean precipitation line spreads north from the Gulf Coast as far as the Ohio River and northeast along the Appalachian Mountains to Pennsylvania. The New England and Middle Atlantic coasts also show a mean rainfall in excess of 4.0 in (102 mm). All the rainfall contours advance west into the Great Plains, though the extreme western and northern sections still receive less than one-inch (25-mm) totals for March. The Pacific states and the western portions of the Intermountain region show distinct decreases from the winter maximums of the previous three months.

The most devastating feature of March weather is the great increase in tornado frequency, the number being almost double that of February. The average number of occurrences is 43, and the maximum reported in any one March reached 124 in 1961. The greatest tornado event ever witnessed in the United States occurred on March 18, 1925, when the Tri-State Tornado remained on the ground over a period of 3.5 hours and killed 695 persons along its 219-mi (352-km) path from southeast Missouri, across southern Illinois, and into southwest Indiana.

Vegetation usually continues dormant during much of March in the northern tier of states, but growth becomes rapid in southern sections, with such spring crops as oats and corn usually appearing above ground. By the middle of the month the temperature rise is normally sufficient to permit corn planting northward to southeast South Carolina, north-central Georgia, central Alabama, and southern Arkansas, and for early seeding in the southern portion of the spring wheat belt. Planting of spring oats begins as far north as the central portions of the country. Fruit trees will blossom throughout much of the South except in the highlands of the Appalachians.

March

1 1914 Heavy wet snow with high winds, "the worst since '88," crippled New York and New Jersey; *Congressional Limited* "lost"; Asbury Park received 24 in (61 cm) snow; New York's barometer dipped to record 28.38 in (96.1 kPa); downed wires and poles disrupted communications and power in New Jersey.

1980 Norfolk, Va., had 13.7-in (35-cm) snowfall to boost its seasonal total to record 41.9 in (106 cm); snow fell in Florida, and Miami registered 32°F (0°C).

2 1846 Great northeast storm and tide in Virginia and Carolinas; record tide 5 ft (1.5 m) above normal at Norfolk; $500,000 damage; 50 families, 1000 cattle reported drowned on Nott's Island, N.C.

3 1966 Deadly tornado hit Jackson, Miss.; 54 killed.

4 1909 President Taft inaugurated in furious storm despite fair weather forecast; about 10 in (25 cm) wet snow disrupted travel and communications; storm of criticism against Weather Bureau followed.

1966 Severe blizzard raged in North Dakota and Minnesota for over 100 hours; winds 100 mi/h (160 km/h) in gusts; snowfalls up to 35 in (90 cm); traffic paralyzed for three days.

5 1872 Severest modern March cold wave in East; New York City had 3°F (−16°C) and Boston −8°F (−22°C) on March 6, "coldest since March 1833."

6 1723 Famous colonial high tide in Massachusetts Bay, unequaled until Lighthouse Storm in 1851. Boston streets inundated. Described by Benjamin Franklin in local press and by the Rev. Cotton Mather in *Transactions of Royal Society*.

1962 Great Atlantic Coast Storm of '62 destroyed over $200 million property from New England to Florida; major shoreline erosion from Long Island to North Carolina; 70-mi/h (113-km/h) winds raised waves 40 ft (12 m) high; deep snows of 33 in (84 cm) fell in Virginia mountains.

7 1932 Severe coastal storm set barometric records from Virginia to New England; lowest at Block Island, 28.20 in (9.5 kPa); winds high, but little erosion occurred.

8 1909 Tornado in Dallas and Monroe counties, Ark., killed 64, injured 671.

9 1862 Historic battle between *Merrimac* and *Monitor* took place under pleasant anticyclonic conditions; Fort Monroe nearby reported 55°F (13°C), fair skies, barometer 30.40 in (102.9 kPa), and light west wind.

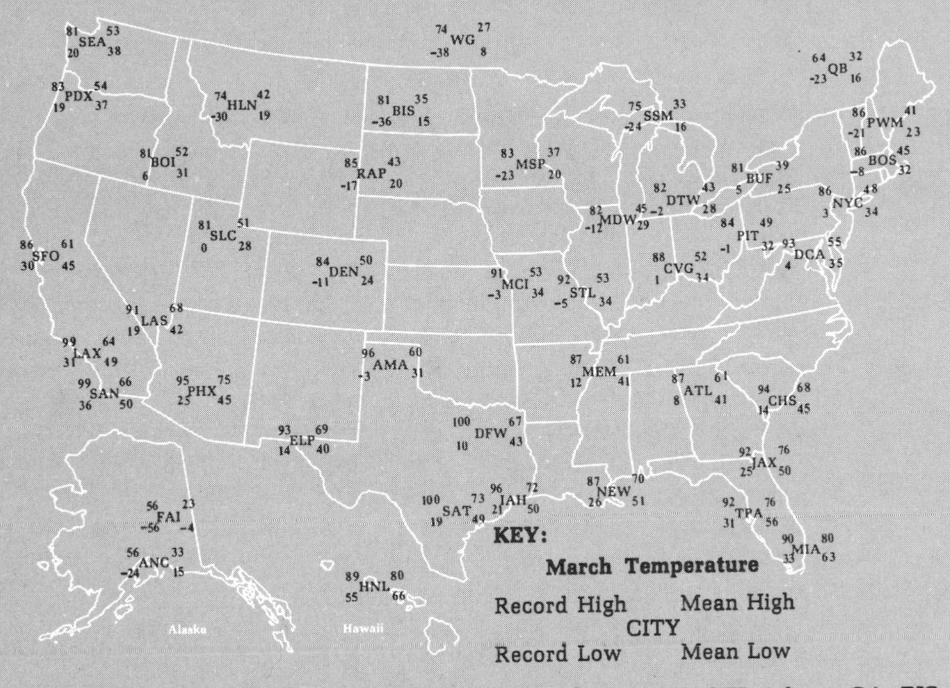

Station Designators: AMA Amarillo TX; **ANC** Anchorage AK; **ATL** Atlanta GA; **BIS** Bismarck ND; **BOI** Boise ID; **BOS** Boston MA; **BUF** Buffalo NY; **CHS** Charleston SC; **CVG** Cincinnati OH; **DCA** Washington DC; **DEN** Denver CO; **DFW** Dallas-Fort Worth TX; **DTW** Detroit MI; **ELP** El Paso TX; **FAI** Fairbanks AK; **HLN** Helena MT; **HNL** Honolulu HI; **IAH** Houston TX; **JAX** Jacksonville FL; **LAS** Las Vegas NV; **LAX** Los Angeles CA; **MCI** Kansas City MO; **MDW** Chicago IL; **MEM** Memphis TN; **MIA** Miami FL; **MSP** Minneapolis-St. Paul MN; **NEW** New Orleans LA; **NYC** New York NY; **OME** Nome AK; **PDX** Portland OR; **PHX** Phoenix AZ; **PIT** Pittsburgh PA; **PWN** Portland ME; **QB** Quebec QUE; **RAP** Rapid City SD; **SAN** San Diego CA; **SAT** San Antonio TX; **SEA** Seattle WA; **SFO** San Francisco CA; **SLC** Salt Lake City UT; **SSM** Sault Ste. Marie MI; **STL** St. Louis MO; **TPA** Tampa FL; **UL** Montreal QUE; **WG** Winnipeg MAN; **YC** Calgary ALB.

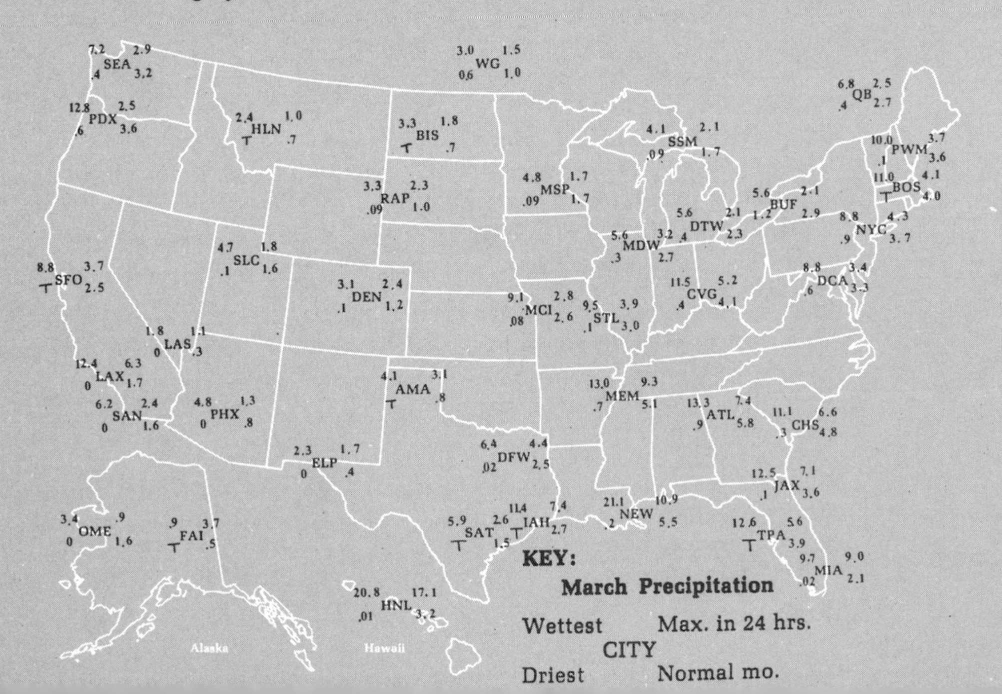

10 1912 Los Angeles reached lowest barometer of record, 29.26 in (99.1 kPa); San Diego 29.46 in (99.8 kPa) also a record.

1960 Heavy snowstorm in Southeast left 10 in (25 cm) in Georgia, 22 in (59 cm) in Tennessee, 24 in (61 cm) in Kentucky, and 15 in (38 cm) in Virginia; severe month of March followed.

11 1962 Record heavy March snowfalls in Iowa; "one of the most paralyzing snowstorms in decades"; 48 in (122 cm) on ground at Inwood after storm.

12 1888 Blizzard of '88, eastern style, dropped 30 in (76 cm) at Albany and 28 in (71 cm) at New Haven in 24 hours.

1967 Great California storm raged for four days, 96 in (244 cm) of snow in 60 hours at Squaw Valley; winds of 90 mi/h (145 km/h) closed mountain passes; lowlands deluged with heavy rains, floods.

13 1888 Blizzard of '88 three-day snow totals: 58 in (147 cm) at Saratoga, N.Y.; 50 in (127 cm) at Middletown, Conn.; 46.7 in (119 cm) at Albany; 44.7 in (114 cm) at New Haven, Conn.; 20.9 in (53 cm) at New York City; 10 in (25 cm) slushy mass at Boston.

14 1870 Severe snow and wind storm struck Iowa and Minnesota; first use of term *blizzard* to describe a severe snowstorm appeared in the *Estherville* (Iowa) *Vindicator*.

15 1941 Severest blizzard of modern era struck on a Saturday night when many were abroad; 39 lost in North Dakota, 32 in Minnesota; 85 mi/h (137 km/h) at Grand Forks, N.D., 75 mi/h (121 km/h) at Duluth, Minn.; front crossed Minnesota in seven hours at 30 mi/h (48 km/h).

16 1942 Two tornadoes, 24 minutes apart, struck Baldwin, Miss., causing 65 deaths.

17 1776 Evacuation Day at Boston. British, frustrated by southerly gales in attempt to dislodge Americans from Dorchester Heights, quit port under favorable wind conditions.

1892 Record heavy snow in Tennessee: 26 in (69 cm) at Riddleton in three-day storm; 18 in (46 cm) at Memphis.

18 1925 Great Tri-State Tornado in Missouri, Illinois, and Indiana; worst U.S. tornado disaster with 695 deaths, including 234 at Murphreesboro and 126 at West Frankfort in Illinois; seven other tornadoes same day raised total deaths to 792.

19 1956 Boston's second heavy snowstorm in three days; 19.5 in (50 cm) at nearby Blue Hill; contributed to snowiest March in area.

20 1948 Juneau had 31-in (79-cm) snowfall in 24 hours, greatest ever at the Alaska capital.

21 1932 Series of Alabama tornadoes killed 269, injured 1874, caused $5 million damage.

1952 Tornado outbreak in Arkansas, Missouri, and Tennessee killed 208, did $14 million damage.

22 1936 Great floods at crest on rivers from Maine to Ohio; Pittsburgh reached 6.1 ft (1.9 m), Harrisburg 3.5 ft (1.1 m), and Hartford 8.6 ft (2.6 m) higher than ever before; 107 lives lost; damage placed at $270 million.

23 1913 Easter Sunday tornado cut 5-mi (8-km) path through Omaha, Neb.; killed 94 persons; did $3.5 million damage.

24 1912 Kansas City had its greatest modern snowstorm, with 25 in (64 cm) in 24 hours.

25 1843 Second great northeast snowstorm spread from Gulf of Mexico to Maine; Natchez, Miss., had 3 in (8 cm) and east Tennessee 15 in (38 cm); contributed to 204-in (518-cm) season total on coastal Maine.

26 1913 Great Miami River Flood in Ohio; crest at Dayton 8 ft (2.4 m) higher than known before; 10 in (254 mm) rain fell over a wide area; more than 467 deaths in Ohio.

27 1890 Louisville, Ky., tornado killed 78 people and did $4 million damage amidst tornado outbreak through Ohio Valley.

28 1920 Chicago's worst tornado disaster; 28 killed; $3 million damage.

29 1921 Washington, D.C., had sharp temperature drop from 85°F (29°C) to 26°F (−3°C), ending early "false spring."

30 1823 Great northeast storm raged with hurricane-force winds from Pennsylvania to Maine; severe in New Jersey; many trees uprooted; high tides along coast; heavy snow fell inland.

31 1962 Florida's worst tornado disaster at Milton in northwest panhandle killed 17, injured 100.

Wind

The Devil sends the wicked wind
To blow our skirts knee-high,
But God is just and sends the dust
To blind man's peering eye.

—Unknown

Wind is defined as the horizontal component of natural air moving close to the surface of the earth. Vertical motions are not included; they are referred to as currents, ascending or descending, or as updrafts or downdrafts. For meteorological purposes, the forward movement of wind is measured as speed, not velocity, since the latter is a vector and implies direction as well. Speed is measured by a rotor or pressure instrument called an anemometer, giving readings in statute miles per hour (mi/h) or nautical miles per hour (knots [kt]), and their metric equivalent. In scientific laboratory work, meters per second (m/s) are generally employed, but in general meteorological work kilometers per hour (km/h) are used. The following are the equivalents:

mi/h	kt	km/h	m/s	ft/s	ft/m
1.0	0.9	1.6	0.4	1.5	88

Wind direction is established by a vane that is positioned aerodynamically by the flow of the wind. The scale is divided into 360 degrees on a full circle. Thus, 90° = east, 180° = south, 270° = west, 360° = north. For marine and popular usage, however, the full circle is subdivided into 32 segments. These provide the main directions called cardinal points: north, east, south, and west. When subdivided again into northeast, southeast, southwest, and northwest, we have the main directions employed in weather observing. Further refinements, such as north-by-east, north-northeast, etc., are employed in marine work.

Winds of damaging strength are defined by the following criteria:

	Mi/h	*Km/h*	*Knots*
Gale	39–54	63–87	34–47
Storm	55–73	88–117	48–63
Hurricane	74+	118+	64+

Highest Surface Wind — Extreme wind speeds in severe tornadoes and mature hurricanes have never been fully measured by a recording device, but they have been estimated to be in the vicinity of 200 mi/h (322 km/h). (See "Hurricanes" and "Tornadoes.") Momentary gusts may exceed that figure, as was reported atop Mt. Washington, New Hampshire (6162 feet [1909 m]) on April 12, 1934, when a mechanical contact anemometer indicated an instantaneous blast at the rate of 231 mi/h (372 km/h). No other measurement at that location has exceeded 180 mi/h (290 km/h) in the past 48 years. The latter figure may be considered the maximum gust to be expected at a United States location where flow is accelerated by a funneling effect over a mountain ridge.

A recent survey of surface winds at the standard 33-ft (10-m) elevation anemometer exposure indicated 130 mi/h (209 km/h) to be the maximum speed to be expected at United States locations.*

Wind gusts of nearly 100 mi/h (161 km/h) have been experienced in the greatest cyclonic storms in non-hurricane and non-tornado situations. The Northeast in November 1950 recorded a maximum gust of 110 mi/h (177 km/h) and in the Northwest the great wind of January 1921 was clocked at 95 mi/h (153 km/h) by standard anemometers. It apparently requires an ascending or descending flow on a mountain slope for extreme speeds. Recently, a descending canyon wind at Boulder, Colorado, hit 143 mi/h (230 km/h). Initial blasts of thunderstorms or line squalls have been measured in excess of 100 mi/h (161 km/h).

Famous Windstorms

Northeast — The Great Appalachian Storm on November 25–26, 1950, achieved the region's greatest sustained wind force when strong gales continued at many locations for 12 hours or more. At coastal stations, such as Newark, New Jersey, and Boston, Massachusetts, speeds sustained for a full minute were in excess of 80 mi/h (129 km/h). Peak momentary gusts were caught at 110 mi/h (177 km/h) at Concord, New Hampshire, 108 mi/h (174 km/h) at Newark, New Jersey, and 100 mi/h (161 km/h) at Hartford, Connecticut. Atop Mt. Washington a wind gust hit 160 mi/h (258 km/h) from the southeast early on November 26. Central Park in the heart of sheltered Manhattan Island set an 80-year record with a fastest mile moving at 70 mi/h (113 km/h).

Lower Lakes — The "Freshwater Fury" on November 9–10, 1913, has been celebrated as the most disastrous in the region's history. At Port Huron, Michigan, the storm raged for more than 24 hours at gale force. The maximum recorded was 49 mi/h (79 km/h) be-

* The four-cup anemometers employed by the Weather Bureau prior to 1928 were found to register speeds too high by 20 percent when indicating 50 miles per hour and 25 percent too high when indicating 100 miles per hour. In this work, speeds prior to 1928 have been corrected to current readings as indicated by standard three-cup anemometers.

tween 6:00 and 8:00 P.M. on the ninth. At Cleveland, Ohio, the wind averaged 40 mi/h (64 km/h) for 16 hours, and Buffalo, New York, had an extreme speed of 62 mi/h (100 km/h). Ten large ore-carriers were lost along with 270 sailors in the storm-tossed waters of Lake Huron.

Upper Lakes — The Armistice Day Storm in 1940 still lives in the memories of many residents of the Upper Midwest as its greatest storm experience. The deep cyclonic center moving across southern Minnesota, western Wisconsin, and Upper Michigan caused a blizzard, the western sector of the storm, that claimed 49 lives in Minnesota. To the east, Lake Superior and Lake Michigan were blasted by whole gales: In Michigan, Alpena, 61 mi/h (98 km/h); Muskegon, 67 mi/h (108 km/h); Escanaba, 49 mi/h (79 km/h); Milwaukee, Wisconsin, 54 mi/h (87 km/h); and Chicago, Illinois, 42 mi/h (68 km/h) — all from south or southwest. A total of 70 sailors lost their lives on Lake Michigan.

Texas Norther — The rush of arctic air from a cold Canadian anticyclone southward over the Great Plains is a spectacular event, sometimes spanning the distance from the Dakotas to Texas in only twelve hours. The norther upon arrival in Texas often brings a quick drop in temperature of 25F (14C) degrees in one hour and as a much as 50F (28C) degrees in three hours. Sometimes it is a wet norther accompanied by driving sleet or snow; at other times a dry norther arrives with clear skies but very strong winds. Amarillo in the northwest panhandle has clocked northers at 70 mi/h (113 km/h), Abilene at 60 mi/h (97 km/h), and Lubbock at 59 mi/h (95 km/h). They may extend their influence to the extreme southern tip of Texas, where Brownsville has endured northers at a 55 mi/h (89 km/h) clip.

Great Plains — Peak winds in severe blizzards on the Great Plains normally mount to the 50 to 60 mi/h (81 to 97 km/h) range in gusts

Opposite: "Galloping Gertie" — the collapse of the Tacoma-Narrows Bridge over an arm of Puget Sound, Washington, November 7, 1940. Courtesy of the Washington State Historical Society.

and may blow steadily close to 40 mi/h (64 km/h) for several hours. A strong blizzard struck the Dakotas on January 6–7, 1903, with the wind at Bismarck reaching 58 mi/h (93 km/h), and Sioux City, Iowa, registered the same speed. Another blizzard on January 29, 1909, set station records at Minneapolis, Minnesota, with 53 mi/h (85 km/h), at Lincoln, Nebraska, with 56 mi/h (90 km/h), and Omaha, Nebraska, with 51 mi/h (82 km/h). Since 1940 with modern wind-measuring equipment, Bismarck, North Dakota, has recorded the following fastest miles of wind passing the station, all from the northwest:

November	67 mi/h (108 km/h)
December	61 mi/h (98 km/h)
January	54 mi/h (87 km/h)
February	54 mi/h (87 km/h)
March	65 mi/h (105 km/h)

Chinook — The most welcome wind in the western portion of the country is the chinook, whose arrival signals the end of a cold spell and relief for both man and beast from the rigors of winter. The chinook is best developed along the eastern slopes of the Rocky Mountains from Alberta to Colorado, but can occur in the lee of any substantial mountain range.

The chinook usually blows from the southwest or west and is dynamically heated by compression in its descent from higher elevations to the plains or valleys. The onset is sudden and may cause a temperature jump of 20 to 40F (11 to 22C) degrees in 15 minutes. After the first rise, the temperature may fluctuate quickly as the locality alternately comes under the influences of strata of either warm air or cold air. The sky is usually clear over the plains, but mountain crests are often covered by a "chinook arch" of clouds. In the Black Hills of South Dakota there have been instances of extreme wind-temperature antics in fluctuating chinook situations. Spearfish, South Dakota, experienced a rise of 49F (27C) degrees in two minutes, and an hour and a half later a plunge of 58F (32C) in 27 minutes. The thermometer had an extreme range from −4°F (−20°C) to 54°F (12°C) during this short period.

The chinook is known as a "snow-eater" because of its ability to evaporate into its dry air a foot or more of snow in a few hours, leaving the ground dry. This opens the range to cattle that may have been deprived of forage for a long period by the snow cover. Charles Russell dramatically caught this scene in his painting *Waiting for a Chinook: The Last of 5000.*

Boulder Canyon Winds — High, gusty winds frequently descend the canyons of the Front Range of the Rocky Mountains in the Boulder, Colorado, area. They occur in the winter half of the year, and on an average of more than once a year. The winds are characterized by extreme gustiness, increasing or decreasing from 10 to over 100 miles per hour (16 to 160 km/h) in a few seconds. An extreme speed of 143 miles per hour (230 km/h) was registered on a gust indicator in January 1971. A study of 76 cases indicated that the most damaging wind is most likely to occur from early evening to early morning and in the month of January. The weather map will show a high-pressure area over western Colorado or eastern Utah with a low-pressure system over the northern Plains. Railroad trains have been blown from the tracks, heavy cranes toppled, and houses demolished by these winds. Much damage is done by wind-blown sand, and small debris is carried through the air like missiles. A windstorm on January 7, 1966, produced over $1 million damage according to insurance claims. Many locations to the lee of the Rocky Mountains experience similar canyon winds, though their characteristics have not been studied as intensively as those near the National Center for Atmospheric Research at Boulder.

Wasatch Canyon Winds — When high pressure is centered over Wyoming and low pressure over Arizona simultaneously, an easterly wind flow over the Wasatch Mountains of northeast Utah descends the western slopes and bursts forth from the canyon mouths into the Ogden–Salt Lake City area. Spring is the favorite time for these visitations. On May 16, 1952, an estimated $1 million damage was caused to the east and north of the Great Salt Lake. There was structural damage to buildings, and many power and telephone lines were downed. Much destruction resulted among parked airplanes. Trees were blown down and orchards devastated. Hill Air Force Base reported a wind of 82 mi/h (132 km/h) with gusts to 95 mi/h (153 km/h).

Other severe windstorms of this type occurred on June 3–4,

1949, and on September 21–22, 1941. During the latter, winds reached gusts of 80 mi/h (128 km/h) at the Salt Lake City Airport, well out on the plain toward the lake. On another occasion, on April 3, 1964, winds at the surface gusted to more than 60 mi/h (97 km/h) at Salt Lake City Airport, Ogden, and Hill Air Force Base. Trucks and trailers were blown from the highway on Route 91, opposite Farmington Canyon. Pilots reported unusual turbulence and the presence of strong atmospheric mountain waves over the mountains.

Northwest — The Great Olympic Blowdown on January 21, 1921, produced the mightiest wind force in the history of the Northwest Coast to that time. It swept the coastal plain from central Oregon to southern British Columbia. A small but violent disturbance moving north parallel to the coast was the cause of the rising wind, which increased within a matter of minutes to about 73 mi/h (117 km/h). The only government anemometer in the peak wind area collapsed at that speed, but the observer estimated the gusts attained 85 mi/h (137 km/h) subsequently. The damaging winds were confined mainly to the coastal strip between the Olympic Mountains and the Pacific Ocean. An estimated eight billion board feet of timber in the heart of one of the finest stands of Douglas fir were destroyed.

The Columbus Day "Big Blow" in 1962 gave the inland sections of western Oregon and Washington almost as severe a wind lashing as did the 1921 storm on the coast. Again, a very deep cyclone of relatively small dimensions moved along the coast. A wind gust of 96 mi/h (155 km/h) was registered at Astoria, at the mouth of the Columbia River. Portland had a fastest mile of 88 mi/h (142 km/h) and the Seattle-Tacoma Airport 65 mi/h (105 km/h). Total damage to crops and structures was estimated at $175 million and 13 people were killed. A storm of somewhat similar character hit the same areas on January 9, 1880, with devastating effect.

Santa Ana Winds — Southern California can be subjected to strong northerly to easterly gales that descend from the interior plateau of the Mojave Desert when high pressure prevails there

Tower of Campbell Hall on Oregon College of Education campus at Monmouth, Oregon, topples under Columbus Day Big Blow in 1962. Note pine trees straining with the wind. Photo by Wes Luchau.

and low pressure lies along the coastal plain. The winds rush through the mountain passes and down to the ocean in San Bernardino, Riverside, Ventura, Los Angeles, Orange, and San Diego counties. The name comes from the Santa Ana Valley, where early American settlers felt the lash of the winds as had their Mexican predecessors.

Originally cool and very dry, the airstreams are heated in their descent and arrive as hot, desiccating blasts causing great physical discomfort. At speeds of 40–50 mi/h (64–81 km/h), they also do structural damage, cause crop losses, and create hazardous fire situations. Their influence is greatest at the outlets of the mountain passes, but sometimes the effect is widespread down valleys to the coastal plain. Downtown Los Angeles once recorded a peak gust of 49 mi/h (79 km/h) on January 12, 1946, during a Santa Ana.

One of the most damaging Santa Anas occurred on November 5–6, 1961, when wind-fanned fires in the Bel Air–Brentwood section of Hollywood spread disaster. A weather observer only 1.2 mi (1.9 km) from the fire area recorded a temperature of 92°F (33°C), a humidity of 4 percent, and winds of 55 mi/h (89 km/h). Downtown Los Angeles had humidities close to or below 20 percent for 36 consecutive hours. At Burbank the reading of 3 percent was the lowest since the station opened in 1932.

Another tragic Santa Ana situation developed on September 25, 1970, when record late-September temperatures seared Southern California for a week. Los Angeles registered 105°F (41°C), and San Diego 97°F (36°C). A nineteen-month drought made brush and buildings tinder dry. Fires broke out in many separate places, creating the worst fire condition in California's history. San Diego County was especially hard hit, with 500 homes and 500 other structures consumed by the flames.

Northers — The Berkeley Hills, east of San Francisco Bay, are also subject to the same type of destructive fires when downslope winds prevail. The city of Berkeley suffered its worst catastrophe on the afternoon of September 17, 1923, when north and northeast winds spread fires from the hills across much of the northeast section of the city. In three short hours, 577 homes were burned to the ground. The fires reached approximately the corner of Hearst and Shattuck avenues at the northwest corner of the University of California campus. Fortunately, no one was killed in the conflagration. Extensive fire prevention methods adopted after the 1923 blaze paid off in September 1970, when threatening fires were contained within a small area.

April

April, April,
Laugh thy girlish laughter;
Then, the moment after,
Weep thy girlish tears.
— William Watson, "April"

When April blows his horn,
It's good for hay and corn.

If it thunders on All Fool's Day
It brings good crops of corn and hay.
— Richard Inwards, Weather Lore (1898)

Oh the lovely fickleness of an April day." The words of W. H. Gibson well express the contrariness of the weather elements this month, often opening in the cold lap of winter and ending in the warm embrace of late spring. The opening day of the baseball season in northern cities may see the diamond covered with snow, yet by the first of May veteran pitchers find enough summer balm to unlimber their arms for a full nine-inning stint on the mound.

The proverbial April showers are a reality on the meteorological calendar. After the general storms of winter with steady precipitation over wide areas, rainfall becomes more localized, and brief showers over limited areas are a regular feature of atmospheric behavior. The reason for the showery conditions lies in the increased heating of the surface of the earth. The sun's direct rays are now falling farther north, and the heat input per unit area has greatly increased from the winter nadir. While the air aloft still remains cold, heating of the air layer nearest the ground causes it to expand and become lighter or less dense than the air above. It becomes unstable and tends to rise, as a hot-air balloon does, through the denser air aloft. In rising, the air cools at a steady rate, at what meteorologists call the adiabatic lapse rate, a decrease of 5.3°F (2.9°C) per each 1000-ft (305-m) ascent.

The rising bubble of air continues to cool until it reaches its dew point; then the moisture in the rising current condenses and forms a cloud. If conditions are favorable, the cloud grows into a shower cloud with rain falling or perhaps into a thunderhead with thunder and lightning and possibly hail. April is usually the first time in the spring season that atmospheric conditions favor this local convective process. Hence, *showers* and *April* have become synonymous.

The Pacific High continues in about the same latitude as in March and extends farther to the west. The Aleutian Low is centered over the extreme eastern end of the island chain adjacent to the Alaskan mainland; mean pressure at the center has risen to 29.75 in (100.8 kPa). The frequency of great oceanic storms in the region diminishes. A new area of low pressure appears on the weather maps over the southwestern United States and northern Mexico, where the increasing solar angle warms the desert lands and creates what meteorologists call a thermal low. It is now an area of increasing heat and dryness, which discourages storm formation.

In the Atlantic Ocean, the Azores–Bermuda High remains in the same relative latitude while expanding its fair-weather regime westward. To the north, the Iceland–Greenland Low fills a broad zone stretching from near the southern tip of Greenland almost to the Norwegian coast. The mean central pressure has risen further to about 29.83 in (101.0 kPa). The North Atlantic Ocean is approaching its least stormy period of the year.

During April the axis of the main westerly jet stream remains quite far south, at about 38°N, causing April to rank among the stormy months on the mainland of North America. Areas of high storm frequency lie in the Intermountain region, the central Great Plains, and off the Middle Atlantic coast. Since thermal contrasts are still great over middle latitudes, in these regions originate some very intense general storms whose clashing fronts stimulate severe local storm activity. The storm tracks of April closely resemble those of March except for the gradual phasing out of activity over the central Pacific Ocean and California and in the Gulf of Mexico. In the Southeast storm formation has shifted east to the offshore waters of the Atlantic Ocean.

The primary path of anticyclones over North America moves eastward to central Canada as the famous Hudson Bay Highs become dominant in influencing weather patterns. By blocking the progress of coastal storms, they cause an easterly flow over New England and the Middle Atlantic region bringing periods of backward spring weather. In the Intermountain region, the prominent anticyclonic activity of the winter season phases out. The path of high pressure areas from northwest Canada, so active in winter, takes a secondary role as anticyclones from the central Pacific Ocean now cross the northwest quarter of the United States. Both anticyclonic tracks from the west converge over South Dakota. From the northern Plains the combined path dips southeast to the Ohio Valley and then curves east-northeast to pass off the central New England coast. This route brings periods of sunny skies and warming temperatures that makes spring such a welcome gift after the duress of winter.

During April the sun's direct rays move ten degrees northward to about 15°N, in the latitude of Guatemala and Honduras. By the end of the month the length of the day in the latitude of 45°N (Eastport–Minneapolis–Portland) will increase by about one hour and 27 minutes. The augmented influx of heat moves the mean 60°F (16°C) isotherm as far north as Cape Hatteras on the Atlantic Coast, to the bootheel of southeast Missouri on the Mississippi River, and to the high plains of Texas just north of El Paso. The mean freezing line has disappeared except in the western mountains, and only locations along the immediate Canadian border remain north of the 40°F (4°C) line. The 70°F (21°C) isotherm appears in the Southwest and pushes northward up the Colorado Valley. The coldest places this month are found in the higher elevations of central Colorado and western Wyoming.

Temperature extremes in April: maximum 118°F (48°C) at Volcano Springs, California, April 25, 1898; minimum −36°F (−38°C) at Eagle Nest, New Mexico, April 5, 1945.

In the eastern United States, April is less wet than March. The area of 4.0 in (102 mm) or more has decreased somewhat in the Southeast, while the 2.0-in (51-mm) or more line has expanded over the Great Plains west to the 100°W meridian, and the one-inch (25-mm) line covers all of Texas except the extreme southwest. Rainfall expectancy in all the Pacific states is down sharply from the March figures, and in the southern part of the region the rainfall season is all but ended. Locations in the Intermountain region have about the same as, or small increases over, March amounts. April is the wettest month of the year at Salt Lake City, while locations on the eastern slopes of the Rocky Mountains have decided increases in April, building toward the May or June maximums.

In April it is usually too cool for vegetation to develop in the extreme northern sections of the country, but in southern and central areas a month of intense farm activity takes place. In the South the planting of commercial crops is nearing completion, with cultivation begun. Cotton planting begins about April 1 in the central parts of the Gulf States and extends to the northern limits of the suitable area by the end of the month. Corn planting begins about the first as far north as southern Virginia, Kentucky, Missouri, and Kansas, and by the end of the month usually as far north as southern Pennsylvania, central Ohio, Indiana, Illinois, and Iowa. Spring wheat planting is completed in Montana and northern North Dakota. In central districts, winter grains, grass, and pastures frequently make considerable growth, while early planted potatoes begin to mature in Texas. Early fruits and berries bloom as far north as Maryland, Virginia, and the Ohio Valley.

April

1 **1960** The satellite Tiros I, launched from Cape Canaveral, was immediate success; had lifetime of 2.5 months; transmitted 19,389 usable photos from above showing distribution of earth's cloud formations.

2 **1975** Severe storm blasted Northeast for three days; hurricane gusts along coast, 140 mi/h (225 km/h) atop Mt. Washington; 24–36 in (61–91 cm) snow in New Hampshire and Maine.

3 **1974** Super Tornado Outbreak with 148 tornadoes in 12 states; 309 fatalities, 5300 injuries; Alabama, Kentucky, and Ohio hardest hit; Xenia, Ohio, devastated.

4 **1804** Large tornado crossed six Georgia counties; at least 11 killed near Augusta.

5 **1936** Tupelo, Miss., tornado cut path 1200 ft (366 m) wide through residential area; path 20 mi (32 km) long; killed 216, injured 700, damage $3 million.

6 **1936** Gainesville, Ga., tornado; 203 killed, 934 injured, $13 million damage.

7 **1857** Late-season freeze; snow fell in every state; Houston, Tex., down to 21°F (−6°C).

8 **1973** Severe snowstorm, high winds, drifts to 16 ft (4 m) closed highways in Iowa; 20.2 in (51 cm) fell at Belle Plaine; 19.2 in (49 cm) at Dubuque (late season records).

9 **1947** Southern Plains Tri-State Tornado; track 221 mi (356 km) from Texas through Oklahoma to Kansas; 169 killed, 980 injured, damage at $9.7 million; forward speed 42 mi/h (68 km/h); Woodward, Okla., chief sufferer with 101 killed.

10 **1877** First of two great coastal storms struck Virginia–Carolina coast; Oregon Inlet widened by 0.75 mi (1.2 km); "entire topography of country is materially altered"; sand dunes at Hatteras flattened.
1979 Outbreak of 23 tornadoes ravaged Red River areas of Texas

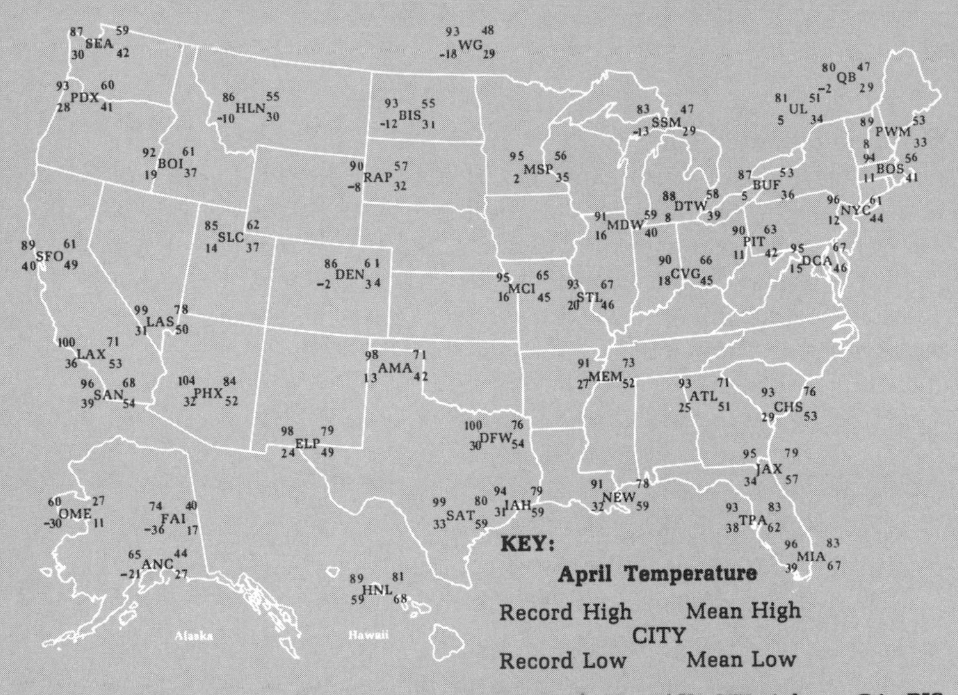

Station Designators: AMA Amarillo TX; **ANC** Anchorage AK; **ATL** Atlanta GA; **BIS** Bismarck ND; **BOI** Boise ID; **BOS** Boston MA; **BUF** Buffalo NY; **CHS** Charleston SC; **CVG** Cincinnati OH; **DCA** Washington DC; **DEN** Denver CO; **DFW** Dallas-Fort Worth TX; **DTW** Detroit MI; **ELP** El Paso TX; **FAI** Fairbanks AK; **HLN** Helena MT; **HNL** Honolulu HI; **IAH** Houston TX; **JAX** Jacksonville FL; **LAS** Las Vegas NV; **LAX** Los Angeles CA; **MCI** Kansas City MO; **MDW** Chicago IL; **MEM** Memphis TN; **MIA** Miami FL; **MSP** Minneapolis-St. Paul MN; **NEW** New Orleans LA; **NYC** New York NY; **OME** Nome AK; **PDX** Portland OR; **PHX** Phoenix AZ; **PIT** Pittsburgh PA; **PWN** Portland ME; **QB** Quebec QUE; **RAP** Rapid City SD; **SAN** San Diego CA; **SAT** San Antonio TX; **SEA** Seattle WA; **SFO** San Francisco CA; **SLC** Salt Lake City UT; **SSM** Sault Ste. Marie MI; **STL** St. Louis MO; **TPA** Tampa FL; **UL** Montreal QUE; **WG** Winnipeg MAN; **YC** Calgary ALB.

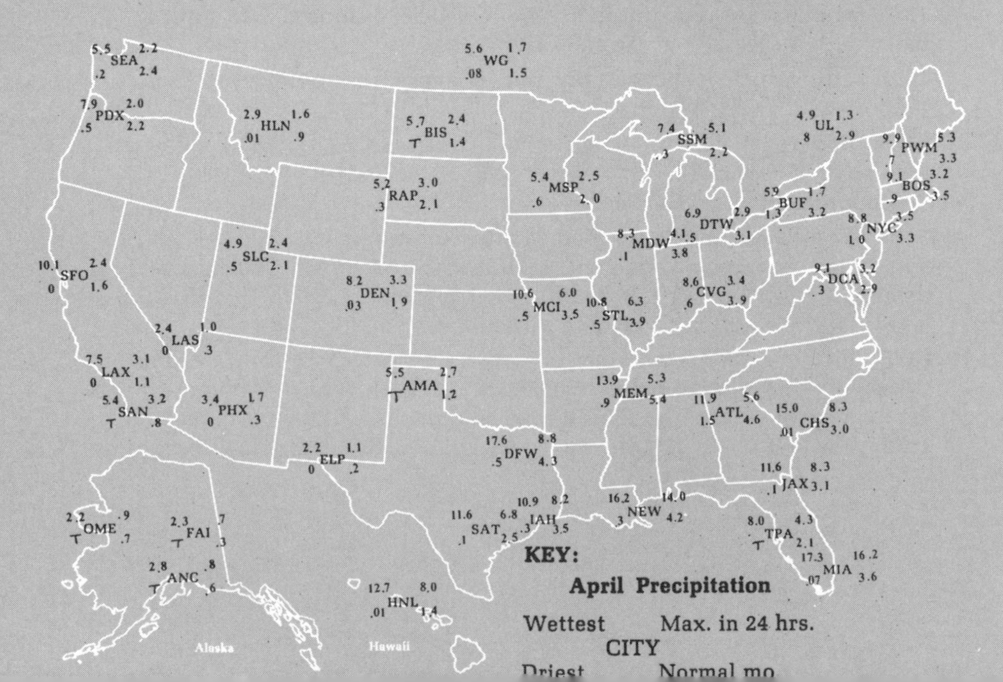

and Oklahoma; Vernon, Tex., had 11 deaths and $27 million damage; Lawton, Okla., 3 deaths; Wichita Falls, Tex., suffered 3095 homes destroyed, $300 million damage, and 42 fatalities.

11 1965 Palm Sunday Tornadoes; 37 funnels descended in Iowa, Wisconsin, Illinois, Michigan, Indiana, and Ohio; 271 killed; 3000 injured; damage placed in excess of $500 million.

12 1861 Bombardment of Fort Sumter in Charleston Harbor; rain all day, 0.55 in (140 mm); 72°F (22°C) at 2:00 P.M.

1927 Tornado wiped out Rock Springs, Tex.; 72 deaths.

1934 Mt. Washington, N.H., registered gust of 231 mi/h (372 km/h); five-minute air flow at average of 186 mi/h (300 km/h).

13 1877 Second severe coastal storm with hurricane-force winds on Outer Banks of North Carolina; more beach erosion and topography transformation.

14 1873 Famous Easter Blizzard in Kansas, Nebraska, and South Dakota; gales blew wet snow into massive drifts; many settlers on land claims perished in three-day storm.

15 1921 Most intense U.S. snowfall: Silver Lake, Colo., at elevation of 10,220 ft (3115 m) measured 76 in (193 cm) in 24 hours, 87 in (221 cm) in 27.5 hours.

16 1851 Famous Lighthouse Storm near Boston Harbor; great tide; whole gales, gigantic waves destroyed Minot Light at Cohasset with two keepers; great shipping loss and coastal erosion.

17 1965 Flood crest at St. Paul, Minn., exceeded previous record by 4 ft (1.2 m); former marks on upper Mississippi generally surpassed downstream to Hannibal, Mo., by May 1; only 12 lives lost due to timely warnings; damage placed at over $100 million.

18 1880 Great tornado swarm over Arkansas and Kansas north to Wisconsin and Michigan; over 24 tornadoes reported; more than 100 killed, 65 at Marshfield, Mo.

19 1775 Lexington-Concord Day had crisp, clear, anticyclonic morning; Prof. Winthrop read his instruments at 6:00 A.M.: 45.7°F (7.6° C), barometer rising, light wind from west, "very fair" sky; he noted: "Battle of Concord will put a stop to observing."

20 1901 Unusually heavy late-season snowfall covered a small area in southeast Ohio: 35.5 in (90 cm) at Warren, 28 in (71 cm) at Green Hill, all in 36 hours.

21 1958 Montana experienced spring snowburst: 55 in (140 cm) at Red Lodge, 61 in (155 cm) at Nye Mine, 72 in (183 cm) at Mystic Lake.

22 1883 Widespread tornado outbreak from Kansas to Louisiana to South Carolina; many funnels; over 200 killed; Beauregard, Miss., destroyed.

23 1885 Denver's greatest snowstorm: 23 in (58 cm) in 24 hours; Idaho Springs, Colo., had 32 in (81 cm).

24 1908 Tornado swarm killed 155 in Mississippi and 37 in Alabama; greatest losses near Hattiesburg, Miss.

25 1875 New York City's latest measurable snowfall: 3 in (8 cm).

26 1834 Killing frost highlighted backward spring in South; severe around Huntsville, Ala.

27 1942 Destructive tornado in Rogers and Mayes counties in Oklahoma; town of Prior hit squarely; 52 killed; damage $2 million.

28 1928 Heavy late-season snow in central Appalachians: Bayard, W. Va., 35 in (90 cm), Somerset, Pa., 31 in (79 cm), Grantsville, Md., 30 in (76 cm).

1973 All-time record crest of Mississippi at St. Louis; 43.3 ft (13.2 m) exceeded former 1844 mark by 1.9 ft (0.6 m).

29 1905 Taylor, Tex., measured 2 in (51 mm) rain in 10 minutes, 2.35 in (60 mm) in 15 minutes.

30 1852 New Harmony, Ind., tornado; track parallel to March 1925 Tri-State Tornado; killed only 16 due to sparse settlement.

The Feather River in Northern California went on a rampage at Christmastime in 1965. Aerial photo shows waters surging through broken levee near Yuba City, December 24. American Red Cross photo.

Floods

Great floods have flown from simple sources.

— Shakespeare, *All's Well That Ends Well*

Rolling, rolling from Arkansas, Kansas, Iowa,
Rolling from Ohio, Wisconsin, Illinois,
Rolling and shouting:
Till, at last it is Mississippi,
The Father of Waters; the matchless great flood
Dyed with the earth of States; with the dust
and the sun and the seed of half the States.

— Stephen Vincent Benét, "Ode to Walt Whitman"

Hundreds of floods, small and great, occur annually in the United States and will continue to occur. And from time to time meteorologic and hydrologic conditions will combine to produce superfloods of unprecedented magnitude. We have every reason to believe that in most rivers past floods may not be an accurate measure of ultimate flood potentialities. It is this superflood with which we are always most concerned. Floods are as much a part of the phenomena of the landscape as are hills and valleys; they are natural features to be lived with, features which require certain adjustments on our part.

— William G. Hoyt and Walter B. Langbein, *Floods* (1955)

A rising flood presents an inexorable force. For a while, dredged channels, reinforced banks, and raised levees may contain the surging waters; but inevitably increasing hydrologic pressures become too great and the lowlands bordering a watercourse are inundated. All the creations of civilization in the area are threatened with destruction, and human life is placed in jeopardy.

Floods arise from many atmospheric situations: Some are caused by long-range developments and others result from quick-breaking events. Many of our greatest inundations develop from excessive rains falling in general winter storms over large river basins following a wet period that has saturated the terrain. When spring snowmelt is added to the current precipitation, enormous runoffs must be handled by the drainage systems. In the north, ice jamming at critical points along a river often leads to serious local flooding during the spring breakup. In the summer and autumn, tropical storms and hurricanes often carry enormous amounts of moisture north to deluge our coastal rivers and estuaries, adding to the height of storm-induced tidal surges. In mountain country in summertime, thunderstorms trigger cloudbursts, creating raging walls of water that descend steep canyons at incredible speeds and scour the hillsides well above normal stream levels — the dangerous flash floods. Local urban flooding may be caused by inadequate run-off facilities or through blocking of natural drainage channels by new developments and construction.

In all, floods take an annual toll of about 200 lives and destroy about $1 billion of property — the figures are rising almost every year.

The Johnstown Flood of 1889

No resident of the Conemaugh Valley suspected the seriousness of the sequence of events set in motion on the quiet holiday afternoon of Thursday, May 30, 1889, when the first drops of a light rain fell over the Allegheny Mountains of west-central Pennsylvania. These were the harbinger of a band of increasing precipitation being generated by a slow-moving low-pressure system approaching from the west by way of the Ohio Valley. Though not of great barometric depth, the narrow weather trough extended in a long finger from Canada to the Gulf of Mexico. It separated cold polar air rushing southward over the Great Lakes region, with temperatures near 40°F (4°C) and accompanied by late-season snow squalls, from warm tropical air streaming northward over the Atlantic coastal plain, having a thermal content near 70°F (21°C) and attended by thunderstorms. Nature set the scene in the clouds for a clash of the elements over Pennsylvania that would produce a rainstorm of great intensity, and man's defective works on earth were to turn an already major flood into a catastrophe of unprecedented proportions.

Johnstown, a manufacturing city of about 30,000 residents, was situated in a narrow river valley on the main line of the Pennsylvania Railroad about one third of the way from Altoona to Pittsburgh. The exact amount of rainfall at Johnstown cannot be determined since the local observer, Mrs. H. M. Ogle, became a victim of the flood. The rainfall has been estimated at 6.2 in (157 mm) by averaging other nearby observations. Measured amounts ranged up to 9.8 in (249 mm) at Wellsboro, Tioga County, in the Susquehanna watershed. It is probable that the higher mountains caught larger amounts. Though of limited east-west extent, the rain front stretched from West Virginia northeast to central New York State, causing high floods in the Potomac, Allegheny, and Susquehanna drainage systems.

On Friday morning, May 31, the steady rain sent the Conemaugh River over its banks into factories, stores, and homes in the narrow valley above Johnstown. When the local rain gauge was carried away by the high water at 10:44 A.M., the river stood 20 ft (6.1 m) above low water; by noon it was "higher than ever known; can't give exact measurement," according to the river reporter.

This was already a record flood and would have achieved historic local stature without the occurrences at South Fork Creek of the Little Conemaugh River, some 15 mi (24 km) upstream from Johnstown. A dam had been finished there in 1852 to supply water for the Pennsylvania Canal. When the canal project was abandoned for the railroad, the "Old Reservoir" remained unused and neglected until a group of wealthy sportsmen acquired the property in 1879 to create a fishing lake and camp. An old break in the South Fork Dam was plugged and the level of the dam lowered two feet to make room for a transverse roadway, which, in turn, necessitated obstruction of the spillway with trestles to support that roadway and the installation of wire-mesh fish guards that impeded the free outward flow of water. Despite these changes, the spillway continued to perform its job of carrying off the discharge of the South Fork in a satisfactory manner.

The dammed stream formed a considerable body of water, measuring over 2 mi (3.2 km) long and a little less than a mile (1.6 km) at the widest point. It covered about 450 acres. When full with spring run-off, it was close to 70 ft (21.3 m) deep. This body of water was impounded by an earthen dam about 900 ft (274 m) long with a crest about 72 ft (22 m) above the stream bed. The surface of the water stood on a mountainside about 450 ft (137 m) above the level of its ultimate rendezvous, the stone bridge in the center of Johnstown.

Johnstown's main business district after the flood. Here stood its greatest stores and its opera house. Irving L. London, Johnstown Flood.

By 11:30 A.M. on the fateful Friday, when flood waters were already raging through the lower part of the valley, the rising level of the heavy run-off impounded by South Fork Dam reached its crest. For three hours the surging waters topped the dam and ate away at its earthen structure. Finally, about 3:00 P.M., the hydraulic pressure became too great. The dam quickly dissolved into a muddy mass and swirled down the steep incline into the valley below. The pent-up waters of the erstwhile fishing lake emptied downstream to join the already swollen Conemaugh.

The event was witnessed by the Reverend G. W. Brown, pastor of the South Fork United Brethren Church*:

* Rev. David J. Beale, *Through the Johnstown Flood by a Survivor,* Edgewood Publishing Co., 1890, 44–45 (no place of publication stated).

> Having heard the rumor that the reservoir was leaking, I went up to see for myself. It was ten minutes of three. When I approached, the water was running over the breast of the dam to a depth of about a foot. The first break in the earthen surface, made a few minutes later, was large enough to admit the passage of a train of cars. When I witnessed this, I exclaimed, "God have mercy on the people below . . ."
>
> The dam melted away . . . Only a few minutes were required to make an opening more than 300 feet wide and down to the bottom. I watched it until the wall that held back the waters was torn away, and the entire lake began to move, and finally, with a tremendous rush that made the hills quake, the vast body of water was poured out into the valley below. Only about 45 minutes were required to precipitate those millions of tons of water upon the unsuspecting inhabitants of the Conemaugh valley. Onward dashed the flood, roaring like a mighty battle, tree-top high, toward South Fork village, rolling over and over again rocks that weighed tons and tons, carrying them a mile or more from the spot where they had lain for ages.

The rush of the waters down the valley of the Conemaugh created a phenomenon characteristic of the rapid movement of a large mass through an atmosphere confined by hills on either side. Most of the survivors said they never saw the wall of water coming, but they heard it. It began as a deep, steady rumble, and then grew louder and louder until its roar was like thunder. An avalanche of snow tumbling down a mountainside creates the same type of noise, and a distinctive aspect of an approaching tornado is a roar. Once described as the noise of a heavy freight train, these natural sounds are now compared to the roar of a large jet plane's takeoff, in which sound waves are generated by swiftly moving objects displacing masses of air.

The few who saw the approaching wall of water were impressed by a cloud of dark spray hanging over the front of the wave. The editor of the local *Tribune* wrote: "The first appearance was like that of a great fire, the dust it raised." One survivor saw "a blur, an advance guard, as it were a mist, like dust that precedes a cavalry charge." Another thought at first there must have been a terrible explosion or fire up the river, "for the water coming down looked like a cloud of the blackest smoke I ever saw." Long remembered by the lucky ones was "the awful mass of spray," later referred to as "the death mist." Much of this, no doubt, was caused by a wind rush out of the front of the advancing wall of water as a large volume of air was displaced and pressured outward. Such a quick movement of air from a descending avalanche has been of sufficient force to snap off substantial trees by the strength of the wind alone.

The rush of water from the South Fork joined with the already flooded mainstream to form a sloping wall of water rising at times 20 to 30 ft (6 to 9 m) high. This proceeded to clean out the valley all the way to Johnstown during the next hour, having traveled at the extraordinary speed for water of 22 ft/s (6.7 m/s) or 15 mi/h (24 km/h). Every tree and telegraph pole within reach of the massive torrent, all the houses, stores, and factories, 50 mi (80 km) of railroad track (ties, rails, and roadbed) with locomotives and trains of cars were dumped in a huge jumble of debris measuring a total of 30 acres (12 ha), all pressed against a stone arch bridge spanning the river in downtown Johnstown. The bridge area became for a time an island of safety for hundreds of humans and animals arriving on floating debris. But the jam soon caught fire, and in the ensuing holocaust many were trapped in the wreckage and burned to death when salvation was near.

Over 2100 valley residents and visitors lost their lives in the unprecedented catastrophe, which evoked worldwide expressions of sympathy and compassion for the destroyed valley and its bereaved survivors.

Some Major Floods in the United States Since 1900

Period	*Rivers or Basins*	*Damage in Millions*	*Deaths*
May–June 1903	Kansas, lower Missouri, and upper Mississippi	$ 40	100
March 1913	Ohio and tributaries	147	467
December 1913	Texas rivers	9	177
January 1916	Southern California	—	22
August 1916	Cabin Creek, Big and Little Coal rivers, West Virginia	5	60

Period	Rivers or Basins	Damage in Millions	Deaths
June 1921	Arkansas River in Colorado	25	120
September 1921	Texas rivers	19	215
Spring 1927	Mississippi Valley	284	313
November 1927	Vermont and other New England states	45	88
May–June 1935	Republican and Kansas	18	110
July 1935	Upper Susquehanna	26	52
March–April 1936	Northeastern States	270	107
January–February 1937	Ohio and lower Mississippi	417	139
March 1938	Southern California	24	79
September 1938	New England States	37	600+*
July 1939	Licking and Kentucky	1	78
April–June 1943	Ohio, upper Mississippi, and Arkansas	172	60
May–July 1947	Missouri and Mississippi	235	29
May–June 1948	Columbia	101	75
June–July 1951	Kansas and Missouri	923	60
April 1952	Red River of North and upper Mississippi	198	11
August 1955	Northeast States	714	186
December 1955	Pacific Coast States	154	61
April–June 1957	Texas, Arkansas, Kansas, Louisiana, Missouri, and Oklahoma	105	18
December 1964	California and Oregon	415	40
March–May 1965	Upper Mississippi, Missouri, and Red North	181	16
June 1965	South Platte	416	16
May 1968	North New Jersey	166	—
January–February 1969	California	399	60
March-April 1969	Upper Midwest	151	—
August 1969	James	116	153
August 1971	New Jersey and Pennsylvania	138	—
June 1972	Black Hills, South Dakota	164	237

* The greater part of the fatalities took place along the coast and in estuaries as a result of tidal surges, but all states except Maine suffered deaths in river floods.

Period	*Rivers or Basins*	*Damage in Millions*	*Deaths*
June 1972	Eastern States	4019	105
May 1973	South Platte	120	—
Spring 1973	Mississippi	1154	33
May 1975	Mississippi	—	—
July 1975	Red River of North	273	4
September 1975	New York and Pennsylvania	296	9
July 1976	Thompson Canyon, Colorado	—	139
September–October 1976	Southern California and Arizona	160	—
April 1977	Kentucky rivers	424	22
July 1977	Johnstown, Pennsylvania	200	76
September 1977	Kansas City creeks	5	23
November 1977	Taccoa, Georgia	—	38
February–March 1978	Los Angeles area	100	20
August 1978	Southeast Texas	100	33
April 1978	Pearl river in Louisiana and Mississippi	1000	15
April 1979	Southeast Texas	500	1
May 1979	Red River of the North	—	—
July 1979	East Texas	750	1
February 1980	Arizona and Southern California	500	36
April 1980	Mississippi and Louisiana	—	—

The Flash Flood Menace

Flash floods have always been a danger to life and property in many sections of the country, especially where small streams in mountainous areas are confined to narrow valleys or canyons. In recent years this type of flood has become the nation's No. 1 weather-related killer, as the figures of over 200 lives lost and property damage in excess of $1 billion in a recent year well demonstrate. The problem has been greatly accentuated recently by "urban development" along rivers and by the increased use of narrow canyons for recreation purposes. Within the present decade a number of tragedies have resulted from man's encroachment on the banks of rivers and streams flowing through narrow valleys.

Rapid City Flood in South Dakota: June 1972

In terms of both lives lost and damage inflicted, the Rapid City Flood on June 9–10, 1972, ranks as the worst natural disaster in the history of South Dakota. Excessive rains of 15 in (381 mm) in five hours resulted when heavy thunderstorm cells stalled to the west and northwest of Rapid City while moist airstreams from a tropical air mass over the Great Plains continued to feed humid air into the storm cells. Much of the intense precipitation fell over the watershed of Rapid Creek, covering less than 100 sq mi (259 sq km), and rushed downstream through a narrow canyon into the city.

Rapid Creek normally passes smoothly and quietly through Canyon Lake, close to downtown Rapid City. But the flood surge brought much debris downstream, which clogged the spillway of Canyon Lake Dam, causing the reservoir pool to rise 12 ft (3.7 m) higher than normal and raise the volume of compounded water to about five times the usual. The dam's failure at 10:45 P.M. released the mass of water and debris to rush downstream, sweeping the residential areas on each side before pouring into the main streets of the downtown section. The arrival of the main flood crest soon after the dam failure contributed to the exceedingly high water marks attained at 12:15 A.M. in the city. The flood reached 6.5 ft (2 m) above the bank-full stage. This huge excess poured into the business district.

In two short hours all the damage was done in Rapid City. Two hundred thirty-six people died, and property damage mounted to over $100 million as a result of the thunderstorm cell's stalling in the hills above the city.

Big Thompson Canyon Flood: July 1976

The last day of July 1976 brought the eve of Colorado's one hundredth birthday, and it fell within the month and year of our nation's bicentennial celebration. It was a Saturday night, when the mountain resorts of the Rocky Mountains were crowded with vacationers seeking a break from the midsummer heat. Since the river flows through a scenic canyon in the first foothills east of Estes Park, the Big Thompson River Valley was filled as usual with a horde of campers occupying almost every available site.

Rescue operations in Montgomery County, Maryland, outside Washington, D.C., following heavy rains and flash floods in July 1975. NOAA photo.

All day Saturday, cool polar air overlay the eastern slopes of the Rockies, while out on the plains moist tropical air held sway. Though the skies were cloudy, no thunderstorm activity developed until 6:00 P.M., when a strong southeasterly flow carried warm, moist airstreams from the plains upslope over the Front Range. Thunderheads quickly developed, with tops towering to 60,000 ft (18,288 m). The storm cells became locked against the mountains over Boulder and Laramie counties, some of the largest remaining stationary over Big Thompson Canyon. Actual total rainfall over the mountains will never be known, but radar estimates ranged as high as 10 in (254 mm) falling within 90 minutes.

Severe flooding began shortly after 7:00 P.M., with the crest reaching and destroying the small town of Drake at the river's junction with its North Fork about 9:00 P.M. Flow rates were extreme, as high as 25 ft/s (7.6 m/s). A wall of water estimated to be 19 ft (6 m) high swept tremendous amounts of debris downstream, including large trees and numerous structures. A survey reported 323 houses and 96 mobile homes destroyed. The Loveland municipal power plant, a brick building, was lifted from its foundations, and much additional damage resulted in the towns on the plains beyond the canyon mouth. Total losses were estimated at $30 million.

The storm took at least 139 lives in its sweep of the canyon. Several more were reported missing long after the flood. Most campers had no warning of the wall of water about to descend on them.

Kansas City Flash Flood: September 1977

A devastating flash flood struck the Kansas City metropolitan area on September 12–13, 1977, as a result of two heavy rains of six hours' duration, both coming within a 36-hour period. Each storm cell dropped 6 to 7 in (152 to 178 mm) generally, and in a restricted area rainfall up to 16 in (406 mm) was measured during the dual storm period. The storms were unique, not only for their huge rainfall totals, but also for their occurrence over the same small area. Each resulted from different meteorological circumstances.

Hardest hit by the rains and subsequent flash flood was the famous Country Club Shopping Plaza along Brush Creek in southwest Kansas City, Missouri, where flood waters surged through its many boutiques and restaurants. Of the 25 deaths reported, 17 were either drivers of or passengers in automobiles caught in the flood.

Dam Failure at Taccoa, Georgia: November 1977

An ill-defined low-pressure area in the Gulf of Mexico on the opening days of November 1977 pushed large amounts of tropical air into the Southeast. The arrival of a cold front from the west triggered excessive downpours over the southern Appalachians on November 5 and 6. Amounts ranged from 5 to 7 in (127 to 178 mm), with the heaviest showers falling on the night of November 5. Small streams rose into high flood. A disaster occurred in northwest Georgia just above Taccoa when a rain-weakened earthen dam gave way. At 1:15 A.M. the released flood waters swept through the campus of Taccoa Falls Bible College, inundating dormitories and student trailer camps. A total of 39 persons were drowned and 45 injured. Property damage to the amount of $1.45 million resulted when trailers, residences, and campus buildings were destroyed.

May

Hail, bounteous May, that dost inspire
Mirth, and youth, and warm desire!
Woods and groves are of thy dressing,
Hill and dale doth boast thy blessing.
Thus, we salute thee with our early song,
And welcome thee, and wish thee long.

— John Milton, "Song: On May Morning"

A wet May
Makes a big load of hay.

A cold May is kindly
And fills the barn finely.

Let all thy joys be as the month of May.

— Richard Inwards, Weather Lore (1898)

"Welcome be thou, faire, fresshe May," are among some of Chaucer's best remembered words. The month seems to have been named after a now obscure Roman deity with the name of Maia, said to be the goddess of spring and growth. Even in ancient times the first day of May was a time for outdoor festivals. In Rome this was the day to honor Flora, the goddess of flowers, whose name would have been a much better designation for the month. The tradition of celebrating May Day continued through the Middle Ages and beyond. The English country people annually erected a maypole and went into the woods that morning to gather "may flowers," or hawthorn blossoms. The girls wore their most attractive dresses, each hoping to be selected as May Queen, much as we now have blossom queens and apple queens in springtime. Unfortunately, our Puritan forefathers frowned upon such pagan frivolities and chastised Thomas Morton for his attempt to transfer maypole activities to Merrymount near Boston in the Puritan domain.

In May the sun mounts higher and higher toward the zenith, and daytime temperatures surge upward. But nights early in the month are still cool and even may be frosty in northern locations. Europeans traditionally considered the last appearance of the forces of the North to occur on the days of the three Ice Saints of May (Mamertus, Pancras, and Servatius), May 11, 12, and 13; thereafter good growing weather was assured. The same generally applies to the northern United States, for tender plants and truck crops are usually safe from frost after May 15, except in mountain areas and notorious frost hollows. Memorial Day was placed on May 30, at the month's close, to ensure that spring flowers would be in bloom to decorate the graves of our military.

The Pacific High commences its seasonal migration to the north in May; the axis of its mean center will be found along 32°N, the latitude running just below the U.S.–Mexican border. The center lies farther west than in the winter months. A northeast lobe of the high extends over the mainland of southern British Columbia and protects that area and the American Northwest from the incursions of Pacific storms much of the month. Meantime, the thermal low over the southern Colorado Valley has developed into a permanent feature of the daily weather maps, with high temperatures and low humidities prevailing.

In the Atlantic Ocean, the Azores–Bermuda High is larger and stronger; its west–east axis lies close to 40°N and its north–south dimensions have increased, extending stable weather conditions over most of the central Atlantic Ocean. The Iceland–Greenland Low retracts somewhat from its April extension; the center is now found south of the tip of Greenland, where the mean pressure reads about 29.83 in (101.0 kPa). As a result of this alignment, May is the most favorable month of the year for ocean and air navigation over North Atlantic routes.

The principal low-level jet stream enters the continent over Baja California as in April, but takes a more northerly course across the eastern United States. The northward build of the Azores–Bermuda High shunts the core of the westerly airstreams over the Ohio Valley, Pennsylvania, and Long Island. Therefore, the southeast quarter of the country enters the driest portion of its year in the absence of general storms.

The geographical progress of the isotherms northward in May is second only to that of April. The mean 60°F (16°C) line advances along the coast north from New York City and extends west, except for a dip south along the crest of the Appalachians, across Ohio to the Chicago area, then west-northwest to southern South Dakota, where it turns southwest to the vicinity of Denver. Finally, it is springtime in the Rockies, where flowers bloom among the dissolving snow banks. Only the higher elevations of New England and the fringe of Lake Superior now average below 50°F (10°C). Ice-out time for the more resisting lakes and ponds usually comes before midmonth at most northern locations. Across the South, the 70°F (21°C) isotherm occupies the territory that the 60°F (16°C) isotherm covered in April. The change of seasons is now most evident: Air-conditioners are put to work in the Deep South, while the flame of the furnace is extinguished in the North. New Orleans adds 313 cooling degree days, and Chicago totals only 208 heating degree days in May.

Temperature extremes in May: maximum 124°F (51°C) at Salton, California, May 27, 1896; minimum −15°F (−26°C) at White Mountain No. 2, California, May 7, 1964.

The principal development in the precipitation pattern is the extension west of the 2.0-in (51-mm) rainfall line all the way across the Great Plains to the eastern slope of the Rocky Mountains. Thundershowers of the convective or frontal type account for the bulk of the rainfall as Gulf of Mexico moisture streams northward over the land in a monsoon-like surge. The movement of the 4.0-in (102-cm) line north up the Mississippi Valley to northern Illinois and Iowa and west into central Kansas and Nebraska also demonstrates the greatly increased moisture borne on southerly winds from the Gulf of Mexico. May and June are the wettest months of the year at Kansas City and Minneapolis. With high pressure building over the Southeast, cyclonic activity diminishes and rain totals decrease except in south Florida. Amounts in the Pacific region become insignificant in California and southern Oregon, while northward along the coast May totals decrease to about 1.5 in (38 mm) at Portland and Seattle.

Farm work is carried on actively during May in all parts of the country, with corn planting beginning during the first ten days of the month even to the northern limits of the corn belt. Sorghum planting is completed in Kansas, Nebraska, and Missouri. Soybeans are set out in the South during the first half of the month, and in the Midwest and Mid-Atlantic region during the second half. May is the month of blossoms for most of the country, and early vegetables will be coming to market for the first time as the month progresses.

May

1 1854 Great New England Flood after 66 hours' steady rain; crest on Connecticut at Hartford rose to 28 ft 10.5 in (8.8 m), highest known to that time.

2 1920 Oklahoma tornado swarm in Rogers, Mayes, and Cherokee counties; 64 killed.

3 1761 Large tornado swept Charleston, S.C., harbor, where British fleet of 40 sails was at anchor; raised wave of 12 ft (3.7 m); many vessels put on beam-ends; four persons drowned.

4 1812 May snowstorm covered ground from Philadelphia to Maine; 12 in (30 cm) near Keene, N.H.; 9 in (23 cm) at Waltham, near Boston.

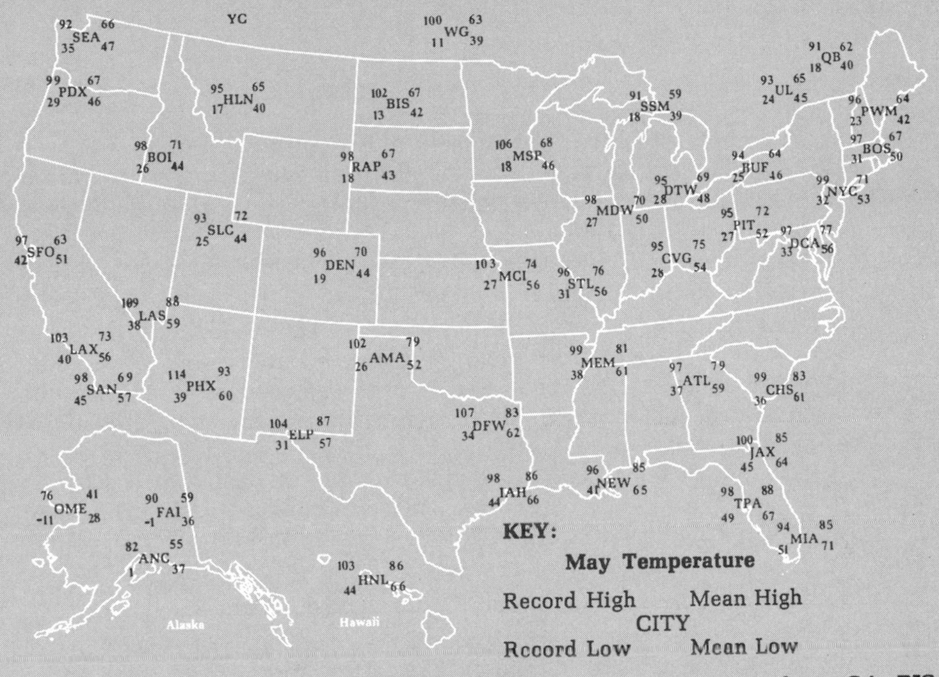

Station Designators: AMA Amarillo TX; **ANC** Anchorage AK; **ATL** Atlanta GA; **BIS** Bismarck ND; **BOI** Boise ID; **BOS** Boston MA; **BUF** Buffalo NY; **CHS** Charleston SC; **CVG** Cincinnati OH; **DCA** Washington DC; **DEN** Denver CO; **DFW** Dallas-Fort Worth TX; **DTW** Detroit MI; **ELP** El Paso TX; **FAI** Fairbanks AK; **HLN** Helena MT; **HNL** Honolulu HI; **IAH** Houston TX; **JAX** Jacksonville FL; **LAS** Las Vegas NV; **LAX** Los Angeles CA; **MCI** Kansas City MO; **MDW** Chicago IL; **MEM** Memphis TN; **MIA** Miami FL; **MSP** Minneapolis-St. Paul MN; **NEW** New Orleans LA; **NYC** New York NY; **OME** Nome AK; **PDX** Portland OR; **PHX** Phoenix AZ; **PIT** Pittsburgh PA; **PWN** Portland ME; **QB** Quebec QUE; **RAP** Rapid City SD; **SAN** San Diego CA; **SAT** San Antonio TX; **SEA** Seattle WA; **SFO** San Francisco CA; **SLC** Salt Lake City UT; **SSM** Sault Ste. Marie MI; **STL** St. Louis MO; **TPA** Tampa FL; **UL** Montreal QUE; **WG** Winnipeg MAN; **YC** Calgary ALB.

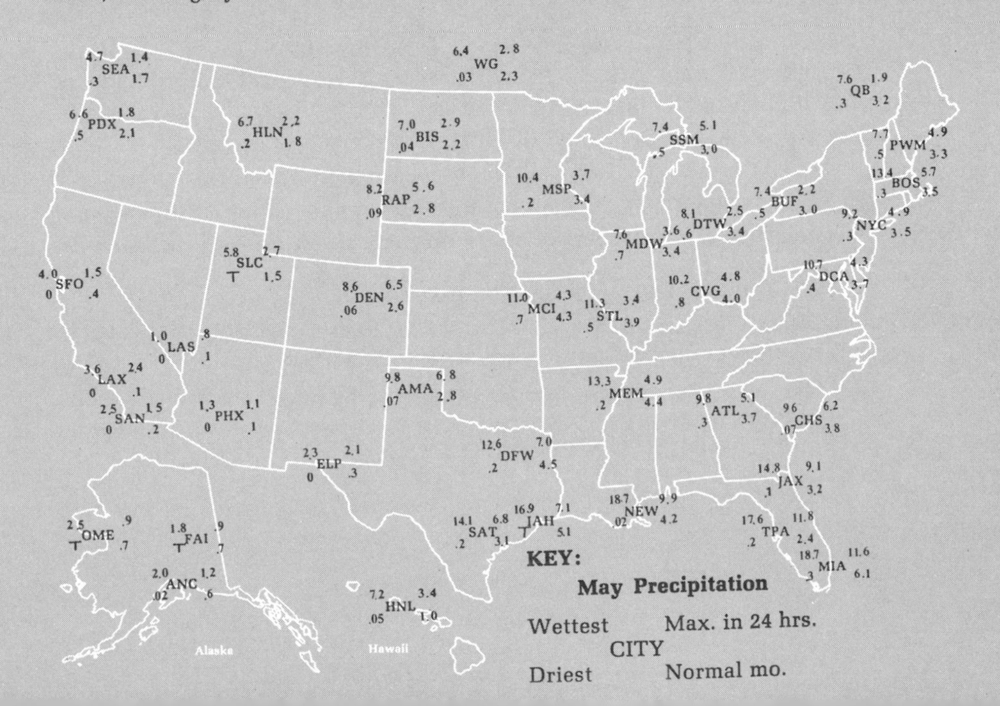

5 1917 Denver's greatest May snowstorm, 12 in (30 cm) falling.

6 1975 Massive tornado struck industrial and residential areas of Omaha; only three killed but property damage of $150 million was most costly in history.

7 1840 Great Natchez Tornado was most deadly and destructive in early U.S. history; city in ruins; 317 killed, mostly along waterfront.

8 1784 Fatal hailstorm at Winnsborough, S.C.; "hailstones or rather pieces of ice measured about 9 inches in circumference; it killed several negroes, a great number of sheep, lambs, geese, and the feathered inhabitants of the woods without number" (*South Carolina Gazette*).

1803 Most famous of all May snowstorms in East spread cover of 6 in (15 cm) and more from southern Indiana to New England; ruined Philadelphia's shade trees; sleighing in Massachusetts.

9 1926 Latest opening of navigation on record at Buffalo, N.Y.

10 1905 Deadly tornado struck Snyder, Okla., killing 87.

11 1970 Large tornado struck downtown section of Lubbock, Tex.; 26 killed; damage $135 million, then a record.

12 1934 Great Dust Bowl storm darkened skies from Oklahoma east to Atlantic Coast.

13 1930 Only authentic death by hail in U.S. Weather Bureau records occurred northwest of Lubbock, Tex., when man was caught in open field.

14 1896 Climax, Colo., registered −10°F (−23°C), lowest ever recorded in United States during May.

15 1834 Greatest May snowstorm of all time in New York and New England; hills around Newbury, Vt., covered with 24–36-in (60–90-cm) snow.

16 1874 Mill Creek disaster west of Northampton, Mass.; dam slippage after rain caused flash flood with loss of 143 lives and $1 million property damage.

17 1794 "Uncommon frost" in all New England destroyed early crops; described in *Collections of the Massachusetts Historical Society*.

18 1825 Burlington Tornado said to have crossed all of Ohio; smashed small log cabin settlement of Burlington, 23 mi (37 km) northeast of Columbus.

1980 Mt. St. Helens in Washington erupted; smoke plume rose to 80,000 ft (24,400 m); ash fell heavily to northeast; cloud reached East Coast in three days; circled world in 19 days.

19 1780 The Dark Day, famous in New England tradition; almost nighttime at noon; chickens went to roost; people fearful of Divine wrath; caused by western forest fires.

20 1957 Tornado track of 71 mi (114 km) ended in Kansas City suburb; 48 dead.

21 1860 Tornado swarm in Ohio Valley hit Louisville, Cincinnati, Chillicothe, and Marietta doing $1 million damage.

22 1832 Kennebec River flood at Waterville, Me.; discharged 140,000 cubic feet per second; high stage not equaled until 1901, exceeded in 1936.

23 1882 Late May snowstorm blanketed eastern Iowa with 6-in (15-cm) depth.

24 1894 Snowstorm across Kentucky of 6 in (15 cm); snow lay all day.

25 1955 Udall Tornado destroyed small Kansas town about 25 mi (40 km) southeast of Wichita, killing 80 persons.

26 1917 Long tornado on tri-state track from Missouri to Indiana; damage trail extended 293 mi (472 km); observed for 7 hours, 20 minutes; struck Mattoon and Charleston in central Illinois; 70 killed.

27 1896 Great St. Louis Tornado; 306 killed; $13 million damage.

28 1887 San Francisco's highest May temperature: 97°F (36°C).

29 1951 Massive hailstorm from Wallace to Kearny counties in Kansas; $6 million damage to crops.

30 1948 Columbia River reached highest stages since 1894; dike broke at Vanport, Ore., flooding area; few of 18,700 residents escaped with more than their clothing.

31 1889 Johnstown Flood in west-central Pennsylvania; heavy rains caused overtopping and failure of reservoir dam; wall of water swept valley below, causing 2100 deaths and immense damage.

Tornadoes

A tornado is Nature's most awesome spectacle, striking terror into all who witness its approach, for there is no refuge above the ground that affords full protection. In a limited area, property destruction can be almost complete when a well-developed funnel sweeps the surface of the earth, and human life is placed in great jeopardy from devastating wind blasts, deadly flying missiles, and collapsing buildings.

The derivation of *tornado* was influenced by the Spanish *tornado,* past participle of *tornar,* to turn, and from *tronada,* thunderstorm.

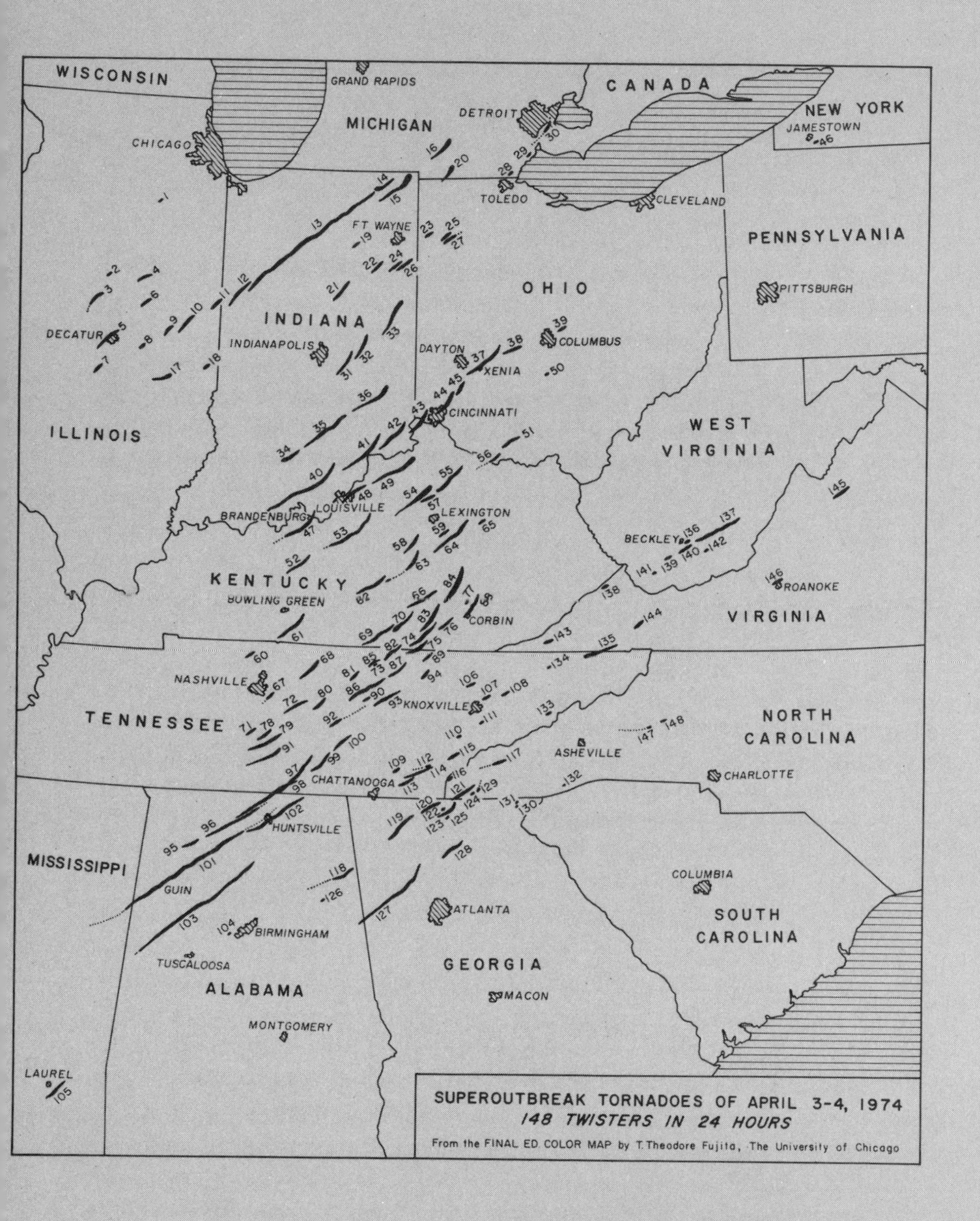

SUPEROUTBREAK TORNADOES OF APRIL 3-4, 1974
148 TWISTERS IN 24 HOURS
From the FINAL ED. COLOR MAP by T. Theodore Fujita, The University of Chicago

There are other less violent local circulations with tornadic characteristics. *Twister* is a popular term, usually reserved for whirls of small dimensions that do minor damage, but its usage is frowned on by meteorologists. Waterspouts have similar, though less violent, tornadic circulations, with added marine aspects. Whirlwinds and dust devils, also relatively small in dimensions and violence, differ in origin in that they derive their energy from direct contact of air with a heated ground surface.

A tornado is defined as a funnel formation that touches the ground, while a funnel cloud has a pendant column aloft without making contact with the ground. Over the past ten years an average of 700 tornadoes have been reported each year; the number is increasing as a result of greater public awareness of the tornado threat and a more efficient reporting system.

The tornado's distinguishing feature is a pendant column narrowing down from a parent cloud, which is almost always a cumulonimbus type associated with a thunderstorm. The visible funnel cloud is formed by the condensation of water vapor around the whirling column, which normally turns counterclockwise, though anticyclonic circulations have been well documented in a few cases. The funnel may assume various forms, from a thin, writhing, gray-white, rope-like appearance to a thick, amorphous, whirling mass of menacing black cloud. Multiple funnels may develop in a mature tornado system, with smaller vortices continually forming and dissipating while whirling around the central core of the tornado.

The funnel or funnels may reach to, or nearly to, the ground. Then great destruction of life and property occurs, and the tornado cloud becomes filled with dust and whirling debris, often obscuring the central funnel after touchdown. The massive destructive power is caused by: (1) blast effect from high wind speed, (2) explosive effect from greatly reduced atmospheric pressure, (3) aerodynamic lifting of roofs and objects through differential pressure, and (4) impact of flying debris and airborne missiles.

Recent studies have shown that the structures of large tornadoes are more complicated than can be seen by the eye. From photogrammetric studies of active tornadoes as well as by site surveys of ground marks afterward, multiple funnels have been discerned, forming and dissipating in rapid succession. These have been labeled subvortices, suction vortices, or suction spots. Tracks of extreme damage have been traced to these lesser funnels imbedded in the larger tornado perimeter, accounting for the varying damage to structures in close proximity.

Classification — A tornado classification scale has been developed to give a numerical representation of a tornado's intensity. It is known as the Fujita-Pearson Tornado Intensity Scale, after Professor T. Theodore Fujita, of the University of Chicago, and Allen Pearson, former director of the Severe Storms Forecast Center of the National Weather Service at Kansas City. It is expressed as three digits: F PL PW.

Professor Fujita devised the intensity scale (F) to provide a means of ranking tornadoes according to the wind speed developed. The F-scale wind speeds are defined as the "fastest ¼ mile of wind." The maximum F scale found within each storm is assigned to it. Photos of typical damage have been supplied to enable one to judge what scale to apply.

The Pearson path-length scale (PL) is based on the actual distance traveled by the tornado, defined as the total length in miles of the storm's path, excluding the portion when the tornado is skipping and not touching the ground. In case the tornadoes form a family or swarm, each member is expressed by a PL scale.

The Pearson path-wide scale (PW) is based on the average width of the tornado damage measured across both sides of the path. The skipped portions are excluded from the averaging, so that the product of the length and the mean width represents the damage area.

The FPP scale is designed for easy assessment of all tornadoes. For general purposes, tornadoes can be divided into three categories: weak (F-0 and F-1), strong (F-2 and F-3), and violent (F-4 and F-5).

A recent survey of 18,545 tornadoes occurring from 1950 to 1977 for which intensity and track lengths could be determined found that 61.7 percent of the annual average of 662 were weak (less that 112 mi/h [180 km/h]), 36 percent were strong (113–206 mi/h [182–332 km/h]), and only 2.3 percent were violent (207–318 mi/h [333–512 km/h]). The violent tornadoes, however, caused 68 percent of the fatalities attributed to the tornadoes for which force estimates could be made.

Ranges of wind speed, path length, and path width corresponding to F, PL, PW

Category	*F (mi/h)*	*PL (mi)*	*PW (yards)*	*Expected Damage*
0	<72	<1.0	<17	Light
1	73–112	1.0–3.1	18–55	Moderate
2	113–157	3.2–9.9	56–175	Considerable
3	158–206	10–31	176–556	Severe
4	207–260	32–99	0.34–0.9*	Devastating
5	261–318	100–315	1.0–3.1*	Incredible
6	319–380	316–999	3.2–9.9*	Inconceivable

Fujita Scale Classification of Tornado Wind Intensity

Fujita Scale	*Name*	*Characteristics of Damage*
F–	Doubtful Tornado (less than 40 mph)	40-mph speed corresponds to Beaufort 8 or "Fresh Gale." Beaufort specification for use on land is "Breaks twigs off trees." Little damage is expected.
F-0	Very Weak Tornado (40–72 mph)	This speed range corresponds to Beaufort 9 through 11. Some damage to chimneys or TV antennas; breaks branches off trees; pushes over shallow-rooted trees; old trees with hollow inside break or fall; signboards damaged.
F-1	Weak Tornado (73–112 mph)	73 mph is the beginning of hurricane wind speed, or Beaufort 12. Peels surface off roofs; windows broken; trailer houses pushed or overturned; trees on soft ground uprooted; some trees snapped; moving autos pushed off the road.

* Miles.

Fujita Scale	*Name*	*Characteristics of Damage*
F-2	Strong Tornado (113–157 mph)	Roofs torn off frame houses leaving strong upright walls standing; weak structures or outbuildings demolished; trailer houses demolished; railroad boxcars pushed over; large trees snapped or uprooted; light-object missiles generated; cars blown off highway; block structures and walls badly damaged.
F-3	Severe Tornado (158–206 mph)	Roofs and some walls torn off well-constructed frame houses; some rural buildings completely demolished or flattened; trains overturned; steel - framed hangar - warehouse–type structures torn; cars lifted off the ground and may roll some distance; most trees in a forest uprooted, snapped, or leveled; block structures often leveled.
F-4	Devastating Tornado (207–260 mph)	Well-constructed frame houses leveled, leaving piles of debris; structures with weak foundation lifted, torn, and blown off some distance; trees debarked by small flying debris; sandy soil eroded and gravel flies in high winds; cars thrown some distance or rolled considerable distance, finally to disintegrate; large missiles generated.
F-5	Incredible Tornado (261–318 mph)	Strong frame houses lifted clear off foundation and carried considerable distance to disintegrate; steel-reinforced concrete structures badly damaged; automobile-sized missiles fly through distance of 100 yds. or more; trees debarked completely; incredible phenomena can occur.

Damage photos illustrating Fujita-Pearson damage scales. Photo by Dr. Theodore Fujita, University of Chicago.

Fujita Scale	*Name*	*Characteristics of Damage*
F-6–12	Inconceivable Tornado (319 mph to sonic speed)	Should a tornado with a maximum wind speed in excess of F-6 occur, the extent and types of damage may not be conceived. A number of missiles such as iceboxes, water heaters, storage tanks, automobiles, etc., will fly through a long distance, creating serious secondary damage on structures. Assessment of tornadoes in these categories is feasible only through detailed survey involving engineering and aerodynamical calculations as well as meteorological models of tornados.

Formation — The exact mechanism causing a tornado to form, or not to form, under similar conditions, has been the subject of increasingly fruitful research in recent years. Tornadoes are usually associated with a thunderstorm cloud. They require a moist airstream at the surface that is warm for the season and usually has a southerly component. A recent survey has shown that 75 percent of all tornadoes are associated with temperatures between 65°F and 84°F (18°C and 29°C) and with dew points greater than 50°F (10°C).

Preceding the tornado development, a temperature inversion aloft with warm air overlying cold air inhibits vertical currents and aids in maintaining the concentration of low-level moisture. At intermediate levels the warm, dry layer usually has a southwesterly air movement, while at high levels a strong westerly flow with jet-stream characteristics prevails.

Often along a front or line of instability, a thunderstorm develops a cyclonic circulation several miles in diameter with lowered pressure and updrafts near the center. The contraction of this tornado cyclone into a tighter circle causes increasing wind speeds leading to a further reduction in pressure and to more violent updrafts. Winds of 50 mi/h (81 km/h) may be raised to 200 mi/h (322 km/h) by this process, known as the conservation of angular momentum, in much the same way that an ice skater with extended arms and legs moves from a slow spin to a fast spin by drawing his extremities into a tight upright position.

Region — There is a seasonal migration of the scene of most frequent tornado occurrence. In winter it lies in the states bordering the Gulf of Mexico; in March and April the major activity advances northward into the central Plains, middle Mississippi Valley, and Ohio Valley, the northern limit usually running from Kansas through Iowa to Ohio; in May and June it encompasses the northern Plains, Upper Midwest, and Great Lakes region and remains potentially active there until September, when a retreat southward commences. By November and December the zone of greatest frequency is once again back in the Gulf States. The central point for tornado occurrence within the lower 48 states lies in south-central Missouri near Grove Spring, some 300 miles from the geographical center of the country, which lies in the extreme north of central Kansas. The difference in the two locations reflects the lesser number of tornadoes observed in the western third of the country and in New England.

A "tornado alley" of greatest activity per latitude extends from north-central Texas, across the central parts of Oklahoma, Kansas, and Nebraska, to the Dakotas, with peak activity between the 97° and 98°W meridians. This general area includes the large cities of Waco, Dallas–Fort Worth, Wichita Falls, Oklahoma City, Enid, and Wichita. A secondary axis of increased tornado activity, curving from southwest to northeast, starts at the Caprock escarpment of west Texas, passes through northwest Missouri, and ends in north-central Indiana.

Another path of tornado frequency, the so-called "Dixie Alley," runs from southeast Texas to northern Alabama, then jumps across the Appalachian Mountains, and re-forms in North Carolina. Other apparent areas of increased activity occur north of Chesapeake Bay, in central New England, in eastern Montana and western North Dakota, and over western Florida.

Oklahoma has the greatest tornado frequency per unit area of any state. More than half of the tornadoes annually occur in the seven states comprising the central and southern Great Plains. Tornadoes are reported annually in almost all the lower 48 states, though they are infrequent in the drier, western third of the country. Funnels and even tornadoes have been sighted in southern Alaska and in the Northwest Territories of Canada, as well as in Hawaii.

Season — Tornadoes may occur at any time of the year, having been reported since 1950 on every date from January 1 to December 31, with the single exception of January 16. They are much less frequent in the autumn and winter months from October through February. January has the lowest monthly average with an expectancy of 14.5 tornadoes. A marked increase takes place in March, with twice as many as February, and April has two and a half times as many as March. The peak month is May, with 143 expected tornadoes, and June has only slightly fewer than May. July and August average fewer than 50 percent as many as May and June.

By April 28, 25 percent of the expected tornadoes of the year have normally occurred. June 4 marks the 50 percent point, and 75 percent have been reported by July 20.

Time of Day — Tornadoes favor the warmer part of the day, when solar heating and thunderstorm development are at a maximum: 60 percent occur between noon and sunset. The median time is about 5:00 P.M., local solar time. The hours of least frequency lie between 3:00 A.M. and 5:00 A.M. There is a geographic variation in time of occurrence. The hourly distribution is more evenly spread throughout the 24 hours for the Gulf States; the Midwest and Plains states have a preference for late afternoon and early evening tornadoes.

Direction of Movement — When the nation as a whole is considered, 87 percent of all tornadoes move from southwest to northeast. Further, an additional 9 percent travel toward the southeast or east-southeast from the northwest quadrant. Tornadoes having any component of motion toward the west are extremely rare. But some have been reported to move from any quadrant, change direction abruptly, follow zigzag courses, become stationary, perform loops, or complete circles.

Speed of Movement — Several tornadoes have been clocked at 60 to 65 mi/h (97 to 105 km/h) along the ground. During a short portion of its course the Great Tri-State Tornado moved at a 73-mi/h (118-km/h) clip. Several have been observed to remain stationary for a number of minutes; a South Dakota tornado hovered in one field for 45 minutes. A slow mover crept along at 5 mi/h (8 km/h) for half an hour, enabling men and animals to outrun it. Average forward speed is about 35 mi/h (56 km/h).

Wind Speeds — Since no wind instrument has ever survived the full impact of a tornado, the determination of wind speeds within a tornado funnel has to be approached indirectly. Observations have been made by Doppler radar and by photogrammetric methods, and engineering surveys of tornado sites have produced estimates of the force required to inflict observed damage. Air movement within a single tornado cloud, of course, varies from second to second, from place to place, and from altitude to altitude. While "considerable damage" can be caused by winds of 125 mi/h (201 km/h), the maximum speeds in a well-developed tornado are placed near 200 mi/h (322 km/h), when "severe damage" occurs. The maximum gusts or effective blast force in an extreme tornado may possibly range up to nearly 300 mi/h (483 km/h), inflicting "incredible damage."

The most destructive force in a massive tornado formation apparently occurs in the subvortices or "suction spots," where the forward speed of the tornado column, the circular movement of the subvortex around the central core, and the revolving spin of the subvortex itself are compounded to extreme velocities. A suction spot has been observed by photogrammetric analysis to possess a movement at the rate of 284 mi/h (457 km/h).

Pressure — Accurate pressure measurements by a recorder directly in the center of a tornado have never been made. A wind of 230 mi/h (370 km/h) would be associated with a pressure drop of at least 1.77 in (6.0 kPa). An aneroid barometer at St. Louis, Missouri, was reported to have registered 26.94 in (91.2 kPa) during a tornado passage. In a near passage at Newton, Kansas, a fall of 0.76 in (2.6 kPa) took place in 10 minutes, followed by a rise of 0.94 in (3.2 kPa) in 17 minutes. Tornadoes are often associated with a "tornado cyclone," or a local area of lowered pressure. A maximum tornado is thought to develop the lowest barometric pressures attainable at the surface of the earth from any type of atmospheric disturbance.

Length of Tracks — The path dimensions of tornadoes vary considerably according to region. Localities east of the Appalachians, in Florida, and west of the Rocky Mountains tend to have shorter path lengths than in the Gulf States, Texas, Great Plains, and Mississippi Valley. A recent survey of 18,545 confirmed tornado occurrences between 1950 and 1977 established an average track length of 6.0 mi (9.7 km), and only two percent had lengths in excess of 30 mi (48.3 km).

In modern times the longest verified track of a single tornado was 293 mi (472 km), in the Mattoon–Charleston Tornado on May 26, 1917; of this total, 188 mi (303 km) were in Illinois and 105 mi (169 km) in Indiana. There was evidence of some skipping action in the early stages.

A path of 240 mi (386 km) was followed in Arkansas on December 21, 1947, and the same distance in Louisiana and Mississippi on November 5, 1948. The Tri-State Woodward Tornado on April 9, 1947, pursued a 221-mi (356-km) path through parts of Texas, Oklahoma, and Kansas. In a study for the period 1916 to 1958, only nine tracks were found to have exceeded 200 mi (322 km).

One of the shortest tracks in official records of the National Weather Service was 45 ft (13.7 m), in a 1954 Wyoming tornado.

Width of Path — The average diameter of tornado tracks varies by region as does the length, and an individual tornado may grow and wane in size during its lifetime. The survey of 18,545 recent tornadoes gave the average path width of 140.8 yd (129 m). Another survey showed half the path lengths were 100 yd (91 m) or less, and only two percent were broader than 700 yd (640 m).

Some large tornadoes have a circulatory system of 1-mi (1.6-km) diameter, and toward the end may expand to whirling cyclonic masses covering 5 to 6 mi (8 to 10 km) before dissipating.

In large tornadoes the greatest violence usually occurs in an area less than 0.5 mi (0.8 km) wide, and frequently in a tight core of about 150 to 200 yd (137 to 182.9 m) wide.

Duration — The longest life of a single tornado funnel was seven hours and twenty minutes along the path of the Mattoon–Charleston Tornado on May 26, 1917, through Illinois and Indiana. The Comanche Tornado on June 3, 1860, spread damage from Cedar Rapids, Iowa, to Van Buren County, Michigan, between 4:00 P.M. and 1:00 A.M. on the fourth, a period of nine hours, but it was not known whether the surface contact was continuous or whether there was more than one tornado.

Often tornado outbreaks occur on successive days as a storm system moves eastward. One of the most devastating instances took place June 8–9, 1953, when the Flint, Michigan, area and the Worcester, Massachusetts, area were hit on successive days. In another striking case on June 18 and 19, 1835, the Midwest experienced tornadoes on the eighteenth and then the Great New Brunswick Tornado in New Jersey and associated whirls descended the next day in the East.

The average life of a tornado on the ground is less than fifteen minutes.

Association with Hurricanes — Tornadoes develop occasionally in a tropical storm circulation, usually ahead of the center in the right front quadrant. Hurricane Beulah in 1967 holds the record for spawning at least 115 minitornadoes, mainly in south-central and coastal Texas. Most of them struck rural districts and caused only minor damage. Some of the inland losses from Hurricane Camille in August 1969 were attributed to tornadoes. A destructive and deadly tornado swept through Charleston, South Carolina, on September 10, 1811, during the approach of a hurricane.

Outbreaks — Tornadoes sometimes come in swarms. The record for a multiple outbreak over a twenty-four-hour period is 148 individual tornadoes in the Superoutbreak on April 3–4, 1974.

Some of the outstanding occurrences of swarms are listed below, with the number of individual tornadoes reported in bold type:

1884, Feb. 19: **60+** est. 800 killed in Miss., Ala., N.C., S.C., Tenn., Ky., and Ind.
1925, Mar. 18: **8** 792 killed in Mo., Ill., Ind., Ky., Tenn., and Ala.
1932, Mar. 21: **27** 268 killed, 1874 injured in Ala., Miss., Ga., and Tenn.
1936, Apr. 5–6: **22** 498 killed in Ark., Ala., Tenn., Ga., and S.C.

Battlefield-type devastation resulted when the Tri-State Tornado swept through DeSoto, Illinois, March 18, 1925. National Archives photo.

1952, Mar. 21–22: **31** 208 killed, 1154 injured in Tenn., Mo., Miss., Ala., and Ky.

1959, May 4: **46** none killed, 2 minor injuries from Okla. to Minn. and Wis.

1965, Apr. 11: **51** 256 killed, over 1500 injured in Iowa, Ill., Ind., Ohio, Mich., and Wis.

1967, Apr. **43** 58 killed, 1068 injured from Mo. to Minn.

1967, Sept. 20–21: **115** 5 killed, 28 injured during Hurricane Beulah in Tex.; most were small minitornadoes.

1974, Apr. 3–4: **148** 315 killed, 5484 injured in Ala., Ga., Ill., Ind., Ky., Mich., Miss., N.C., Ohio, Tenn., Va., and W. Va.

1975, Jan. 9–10: **42** 11 killed, 287 injured in Tex. to Ala., and Okla. to Ind.

1980, Apr. 7–8, **60** 4 killed, many injured from Tex. and Fla. to Wis., Mich., Ohio.

Most Deadly — The Great Tri-State Tornado on March 18, 1925, killed a total of 695 persons in Missouri, Illinois, and Indiana. Including the Tri-State Tornado that day, eight separate tornadoes struck more localities in Kentucky, Tennessee, and Alabama, causing a total of 792 deaths, the most tragic tornado day in our history.

The tornado swarm spreading from the Southeast to the Ohio Valley on February 19, 1884, killed an estimated 400 to 800, although there was no means of securing an accurate count. The Superoutbreak on April 3–4, 1974, killed 309 in twelve states and injured 5530 people. The greatest tornado disaster in early America occurred at Natchez, Mississippi, on May 7, 1840, when 317 were killed, mainly on river barges tied up at wharves, by a single tornado.

The number of people killed by tornadoes in the United States averaged 117 annually for the period from 1916 to 1977, and for the period 1950 to 1977, 112. The year with the fewest fatalities was 1972, with a total of only 27, while the year 1925 easily topped the list with 792 killed.

Five Most Deadly Single Tornadoes, 1950–1978

Date	*Location*	*Deaths*
1953, June 8	Flint, Mich.	116
1953, May 11	Waco, Tex.	114
1953, June 9	Worcester, Mass.	90
1955, May 25	Udall, Kan.	80
1971, Feb. 21	Pugh City, Miss.	58

Five Outstanding Tornadoes

Five of the outstanding tornado disasters of our country are described in some detail below. They were selected because they represent the greatest of their type in different regions of the country. The Natchez Tornado in May 1840 still stands as the greatest natural human tragedy of the Deep South. The Great Outbreak of 1884 probably took a higher toll in human lives than any other group of tornadoes. The Great Tri-State Tornado in 1925 in Missouri, Illinois, and Indiana was the greatest tornado physically ever to develop within the United States and took more lives than any other single funnel. The Worcester County Tornado in 1953 was the worst ever experienced in the Northeast and demonstrates what a large tornado can do anywhere in the country if it strikes a populated area.

The Great Natchez Tornado in May 1840 — The greatest American storm disaster in terms of death and destruction during the entire pre–Civil War period occurred in southwestern Mississippi when a massive tornado smashed through the active port of Natchez Landing and then struck the prosperous commercial and residential city on the bluff above the river. This occurred on the early afternoon of May 7, 1840, when Mississippians were sitting down to their dinner. The death toll from contemporary press accounts amounted to 317 persons, and conservative estimates of the damage to buildings ran higher than $1 million, and when losses of merchandise and personal property were added, the figure mounted to $5 million, immense sums considering the dollar value of the day. These figures far exceeded any other tornado or hurricane loss during the early period of American history.

Henry Tooley, who maintained weather records at Natchez for many years, was a surviving witness and left a vivid account as a trained observer of the tornado's approach and impact:

> The roar and commotion of the storm grew more loud and terrific, attended with incessant corruscations and flashes of forked lightning. As the storm approached nearer, the wind shifted to E. 7 [east at force 7, or about 60 mi/h]. At 1:45 P.M. the storm cloud assumed an almost pitchy blackness, curling, rushing, roaring above, below a lurid yellow dashing upward and rapidly approaching, striking the Mississippi some six or seven miles below the city, spreading desolation upon each side, the western side being the center of the annulus.* At this time a blackness of darkness overspread the heavens, and when the annulus approached the city, the wind suddenly veered to S.E. 8 [southeast force 8, or about 70 mi/h], attended with such crashing thunder as shook the

* *Annulus* — the term employed by James P. Espy, a pioneer American meteorologist, to describe the central core of a cyclonic or tornadic storm.

> solid earth. At 2 the tornado burst upon the city, dashing diagonally through it with such murky darkness, roaring and crashing, that the citizens saw not, heard not, knew not the wide-wasting destruction around them. The rush of the tornado occupied a space of time not exceeding five minutes, the destructive blast not more than a few seconds. At this moment the barometer fell to 29.37 inches.

The storm struck a devastating blow at Natchez Landing, where from fifty to sixty flatboats were tied up. Only six survived the wind and storm wave, which rose to the height of six to eight feet. An estimated one hundred boatmen were drowned. The steamboats *Hinds, Prairie,* and *St. Lawrence* were destroyed and sunk at the Landing, and the Vidalia ferry was caught in midstream. Many died on these vessels.

The destruction in the city itself was described in the *Mississippi Free Trader* next day:

> In the upper city, or Natchez on the hill, scarcely a house escaped damage or utter ruin. The Presbyterian and Methodist churches have their towers thrown down, their roofs broken and walls shattered. The Episcopal church is much injured in its roof. Parker's great Southern Exchange is level with the dust. Great damage has been done to the City Hotel and the Mansion House, both being unroofed, and the upper stories broken in. The house of sheriff Izod has not a timber standing, and hundreds of other dwellings are nearly in the same situation. The Court House at Vidalia, parish of Concordia [Louisiana], is utterly torn down, also the dwelling houses of Dr. M'Whortor and of Messrs. Dunlap and Stacey, Esqrs. — The parish jail is partly torn down.

This massive tornado was reported to have struck Natchez Landing and Vidalia, on the opposite side of the river, simultaneously at a point where the waterway was 0.7 mi (1.1 km) wide. The destruction reached in a south-southwest direction about 10 mi (16 km) down the river, and plantations to the north-northeast of the city were also in the path of destruction.

The *Free Trader* concluded: "Our beautiful city is shattered as if it had been stormed by all the cannon of Austerlitz. Our delightful China trees are all torn up. We are peeled [made bald] and desolate."

The Great Southern Tornado Outbreak in 1884 —

> On February 19th, 1884, the States of Virginia, North Carolina, South Carolina, Georgia, Alabama, Mississippi, Tennessee, and Kentucky were visited with the most terrible devastation by wind ever experienced in this country. From 10 o'clock in the morning until 12 midnight sixty tornadoes occurred in different parts of the above-named States. Rough estimates placed the loss of property at from $3,000,000 to $4,000,000; the loss of life at 800, and the number of wounded at 2,500. The number of people rendered homeless and destitute numbered from 10,000 to 15,000, many of whom were left in a starving condition. The number of buildings destroyed was about 10,000. Cattle, horses, hogs, and other domestic animals were destroyed in great numbers. — John Park Finley, *Tornadoes: What They Are* (1887)

The Great Tornado Outbreak of February 1884 stood in a class by itself for number of individual tornadoes, geographic extent of severe turbulence, and number of fatalities until the recent occurrence of the Superoutbreak Tornadoes in April 1974. Yet knowledge and appreciation of the immensity of the 1884 event received scant mention in standard chronologies and listings of major tornado occurrences within the United States. The possibility of a repetition of the scope of the 1884 outbreak was not in the conscious thinking of meteorologists until the 1974 Superoutbreak produced tragedy on the same scale.

Despite the fact that no urban areas were struck by the sixty or more tornadoes, the estimated death toll of 800 has been approached by only one other tornado situation. The swarm of eight tornadoes occurring on March 18, 1925, took 792 lives when the great Tri-State Tornado and others struck several sizable cities and towns in Illinois and Indiana and rural areas elsewhere. The victims of the 1884 tornadoes lived mainly on tenant farms or small plantations, and most were black farm hands and their families. Statistic gathering among such rural populations at this stage of our history was not systematized, so the figures can be employed only as a general estimate.

The great mass of the outbreak occurred in the Southeastern states from central Mississippi (about noon) through Alabama, Georgia, extreme southeast Tennessee, South Carolina, North Carolina, and into southeast Virginia (about midnight). The *Atlanta Constitution,* reporting the event in detail, detected two swaths through Georgia where the principal damage and suffering occurred. One originated in Alabama, cut across the northeast corner of Georgia, and continued through the Carolinas to a point near Cape Henry, Virginia. A second swath farther south crossed Georgia diagonally from its southwest corner to a point on the Savannah River below Augusta and then continued through both Carolinas to an ending near Cape Hatteras.

Two lesser paths cut through Tennessee and Kentucky: one from west-central Tennessee near Dickson and Crossville to east-central Kentucky near Winchester; the other running east-northeast, close to the Ohio River from the vicinity of Paducah, Kentucky, and Metropolis, Illinois, to localities near Evansville, Indiana, and Louisville, Kentucky. Neither the damage nor the size of these tornadoes approached the proportions of those in Georgia or the Carolinas on the same day, and the tornadoes were minor in comparison to the scale of the outbreak in 1974 in the same Ohio Valley area.

A vivid description of a North Carolina funnel in an area near Rockingham, close to the South Carolina border, was published in the *Wilmington Star* soon afterward:

> The center of the storm struck the outskirts of Rockingham with such fury that people were unable to escape from their houses. Buildings were blown into fragments. Some bodies were found under the timbers, others were carried by the wind 150 to 200 yards. A woman was found clasping to her breast an infant scarcely a month old; both were dead. The bodies of victims were terribly bruised and cut, presenting a ghastly appearance. The force of the wind was such that two millstones were moved 100 feet. Chickens and birds were picked clean, except the feathers on their heads. The largest trees were uprooted and smaller ones had all the bark stripped from their trunks. The storm made its appearance at 7:30 P.M., coming from a southwesterly direction from Hamlet, Richmond Co., N.C. The eastern sky was overshadowed by dark flying clouds, tinged with red, growing thicker every minute and the red tinge assuming the hue of fire. At 8:30 there was a heavy fall of rain and hail, the heaviest of the clouds moving westward. At midnight the sky was a dazzling red, and at 1 A.M. there was another heavy fall of rain.

The Great Tri-State Tornado in 1925 — The greatest physical force of a tornadic nature ever to develop within the United States swept relentlessly east-northeast through the heartland of the Mississippi Valley for a period of three and a half hours on the afternoon of March 18, 1925. Its destructive power was exerted mainly over rural farmland, though an occasional village, town, or city lay in the path of oncoming tragedy. The massive funnel made contact with the surface of the earth over a continuous path of 219 mi (352 km), averaging a quarter to half a mile (0.4 to 0.8 km) in width, but sometimes extending its destructive sweep to over a mile (1.6 km). The gathering whirl of wind, cloud, and debris originated in southeast Missouri over rural farmland near Redford, crossed the Mississippi River about 75 mi (121 km) southeast of St. Louis, then raced at a 60-mi/h (100-km/h) clip across 90 mi (145 km) of southern Illinois before dissipating in southwestern Indiana. The Tri-State Tornado ranks first among individual American tornado formations in dimensions, in total area of destruction, and in number of victims.

The heavy toll in deaths and injuries resulted from the tornado's path paralleling a network of railroads in southern Illinois and southwestern Indiana, where it smashed through a series of commercial towns and small industrial cities with population concentrations. The places struck and the number killed in each were: in Missouri — Annapolis 4 and Biehle 4; in Illinois — Gorham 37, Murphysboro 234, De Soto 69, and West Frankfort 148; in Indiana — Griffin 25 and Princeton 45. In addition, about 129 people were killed in rural areas of farmlands and crossroad settlements. An adjusted survey by Red Cross officials released a year later indicated that 695 people were killed in the three states, 2027 injured, and the losses in 1925 dollars amounted to $16.5 million.

The significant meteorological aspects of the Tri-State Tornado include the following: (1) the longest continuous track on the ground, (2) the third fastest forward speed, (3) a continuous exertion of extreme energy, causing severe damage throughout most of its life span, (4) a record 3.5-hour duration on the ground, and (5) a unique location near the center of an advancing low-pressure center that sustained the tornado's energy for an unusually long time.

The thunderstorm responsible for spawning the tornado funnel originated in the southwest sector of a deep low-pressure system, barometer 29.20 in (98.9 kPa), whose center was crossing the Mississippi River into Illinois. The strong cyclonic circulation prevailing provided a favorable environment for the development of thunderstorms and the subsequent formation of the funnel and growth of the tornado. The funnel first touched down west of Redford, Missouri, about 1:00 P.M. It moved at a speed of about 67 mi/h (108 km/h) for the main part of the 85-mile (137-km) track in Missouri and crossed the Mississippi River at about 2:20 P.M. It slowed its forward progress somewhat, to about 60 mi/h (97 km/h), when passing over a series of towns and small cities in southern Illinois.

The tornadic whirl caught up with the parent low-pressure system center about 4:00 P.M., just before crossing into Indiana. Both the tornadic low and the cyclonic low reached their greatest development at this time; central barometric pressure in the cyclone was estimated at 29.00 in (98.2 kPa), but the tornadic central pressure could not be determined. About 30 minutes later (at 4:30 P.M.), the tornadic low either dissipated or lifted from the ground after a 36-mi (60-km) intrusion into Indiana. The average speed along the entire 219-mi (352-km) track was 62 mi/h (100 km/h).

The path was exceptionally straight along the first 183-mi (295-km) course, north 69° east, or east-northeast. The direction did not deviate until the last 20 mi (32 km), when it changed a little more to the north, north 60° east. At the end it was traveling at its highest calculated speed of 73 mi/h (117 km/h), and the width of destruction diminished from a maximum of 1.25 mi (2 km) to about 0.25 mi (0.4 km) in the closing minutes. The total area of severe destruction across the three-state region covered 164 sq mi (425 sq km), equal to a square land area measuring slightly under 13 × 13 mi (21 × 21 km).

The excessive death toll can be attributed to two main factors, aside from the massive destructive power of the giant tornado itself. First, no forecasting system or any reliable communication system existed to warn of the possibility of a tornado and to alert people of the approaching danger. Second, the death-dealing funnel was imbedded in a dark, amorphous mass of whirling clouds and assorted debris that prevented a sighting of the approaching calamity. Many people thought only a severe thunderstorm was bearing down on them rather than a destructive tornado funnel.

If our modern severe-storm warning system had been in operation and our programs teaching how best to cope with the hazard had been instituted, the death toll in 1925 might have been considerably reduced. But the massive size of the Tri-State funnel and its almost complete destruction of property serve as an example of the possible fate of any modern community if it were to be struck squarely by a tornado of such force.

The Worcester County Tornado in Massachusetts in 1953 — Killer tornadoes of massive size have been infrequent in the corner of the country east of the Appalachian Mountains and north of the Carolinas. Though many small tornadoes were reported in colonial times, none seems to have struck a populous area. A tornado swarm in August 1787 covered four New England states, doing much damage but causing only one or two fatalities. The Great New Hampshire Whirlwind in September 1821 was a tornado of massive size, but moved mainly over uninhabited mountain country. So did the Adirondack Tornado in September 1845 in passing from Lake Ontario to Lake Champlain. The Wallingford Tornado in August 1878 in central Connecticut was the first to strike a settled community; thirty-four people died in a brief moment or two. This remained the Northeast's most spectacular tornadic experience for many years, until just after the midway mark of the present century, when both the massive size of the New Hampshire Whirlwind and the high toll of the Wallingford Tornado were surpassed in central Massachusetts.

On the late afternoon of June 9, 1953, a massive tornado formation spread a path of death and destruction for a period of 75 minutes over the rural country and through the towns and cities of Worcester County. Yet its coming was unheralded, and the exact nature of the striking force was unknown to most of its victims. There was no public alert from weather forecasters of the possibility of a tornado, and, even after one had been spawned and commenced its destructive work, there were no warnings to communities ahead of the coming disaster until near the end. The weather bureau forecast issued at 11:30 that morning called for "Windy, partly cloudy, hot and humid, with thunderstorms, some locally severe, developing this afternoon." Some consideration had been given by the forecasters to the possibility of a tornado, in view of the occurrence of tornadoes the previous day in Michigan, but it was felt that "it can't happen here." Even if a tornado alert or community warning had been issued, New Englanders were not "tornado conscious" in 1953, just as they had not been "hurricane conscious" in 1938. The situation has now changed in view of the experiences with these destructive storms.

The Worcester County Tornado funnel first made contact with the ground a short distance west of Petersham about 4:25 P.M., E.D.T. Moving east through Petersham, then southeast, the tornado hit Rutland squarely along Main Street about 4:45 P.M. The track continued southeast parallel to Route 22 to Holden, where for the first time a heavily built-up and populated area came within its destructive sweep. Holden was hit at 5:05 P.M., and soon thereafter the track cut across the northern extremity of the city of Worcester and continued southeast into Shrewsbury. Here in the approximately ten miles from the center of Holden to the center of Shrewsbury the greatest damage occurred and the most people were killed. The tornado's width expanded to almost a mile, with the southern edge of the damage zone about three miles north of the green in downtown Worcester.

The tornado entered Shrewsbury about 5:18 P.M., missing the center of the business area by a fortunate, last-second swerve. After crossing the Worcester Turnpike (Route 9) at the junction with South Street (Route 140), it swerved again and took an eastward track. Westboro was struck a damaging blow at 5:30 P.M. Then the path curved northeast, again crossing the Worcester Turnpike and running parallel to the present route of the Massachusetts Turnpike (Interstate 90). The little community of Fayville was smashed at 5:40 P.M., and then the path of destruction ceased, close to the southeast corner of Sudbury Reservoir, within easy sight of the Framingham Exchange on Route 90.

A witness of the funnel's passage about a half-mile distant

recalled: "It was big and black, coming to a point at the bottom. It was whirling and swirling around." Others reported seeing a vortex shaped like "a snake," "an ice cream cone in the sky," and a "whirling cloud." A press photographer at Indian Lake was able to catch the funnel from a distance of about two miles when the vortex was entering Worcester. His exposure, the only one known to have been taken by a professional, is apparently a good representation of what people observed. A policeman saw the tornado sweeping into Shrewsbury: "It looked like a huge cone of black smoke. Almost like a fire. There was a roar like an express train. Then the wind struck. Rain and hail beat down. There was lots of lightning." Near the end of the trail a Framingham man described the approaching storm: "It looked like heavy smoke and fire coming toward me."

The color of the vortex cloud was generally reported as "black" at a distance of a mile or two, but "brown" at close range with dirt, boards, paper, branches, and miscellaneous debris clearly visible in the whirl.

The approach and passage of the tornado was accompanied by a high-pitched noise described by some witnesses as like "several steam or diesel locomotives" and by others as like "a flight of jet planes." At impact, the pitch crescendoed to a "scream." A Worcester woman remembered: "I began hearing a horrible sound, a kind of whistling and groaning in the sky — something you can't forget. It got louder and more piercing."

From a cross section of 50 persons who had been injured, interviews revealed that 22 had seen the cloud approaching, but only 14 of that number recognized it as a tornado. Not one of the 50 recognized the large hailstones as sometimes being a forerunner of a tornado. None of the 50 had received any official warning by radio, telephone, or police loudspeaker. Thirty-nine recognized some intimation of danger before the personal impact of the tornado, and 28 were able to take some protective measures.

There were three major funnels in central New England that

Opposite: Classic Great Plains tornado funnel in mature stage. Near Cheyenne in western Oklahoma, May 4, 1961. This type is visible for miles across the prairies. Photo by Bill Males.

afternoon. While the Worcester County funnel was losing energy, a tornado swept through Exeter, New Hampshire, doing some structural damage but causing no injuries or fatalities. A third formed to the south of the main Worcester County funnel about 5:30 P.M. and pursued a 29-mile course during the next hour from Sutton to Mansfield. The three storms combined to take 94 lives, inflict over 400 major and 900 minor injuries, and destroy an estimated $53 million in property. At that time this damage figure was the highest for any tornadic occurrence in all American history.

The Superoutbreak of Tornadoes in 1974 — In the space of twelve hours on the afternoon and evening of April 3, 1974, tornadoes swarmed in greater numbers and struck in diverse places with greater total force than ever before in this century. Individual tornadoes have killed greater numbers and inflicted more concentrated damage, but never have death and destruction been spread so widely over our country. Eleven states witnessed funnels dropping from the sky that day, and ten states received sufficient damage to be declared disaster areas. Only the vast outbreak on February 19, 1884, whose statistics remain incomplete, ever approached this disaster in material damage over such a wide area.

By midnight of April 3, when all but minor activity had subsided, a total of 303 people who had seen the sun rise in a threatening sky that morning were dead, and several score communities were a mass of wreckage. Hospitals were crowded with some 5400 injured persons. Those fortunates who had survived uninjured were in a state of shock, grieving for lost family members or close neighbors, and were contemplating the devastation to their homes, schools, and business places. Yet some were also exchanging bits of good news to the effect that a hospital in the path of the tornado had been spared by a last-minute swerve, an empty school had been demolished where a few minutes before dismissal there had been classrooms full of pupils.

At dawn of April 3, a cyclonic circulation aloft over the central Great Plains extended well above 25,000 ft (7620 m), and a strong jet stream had propagated eastward into Oklahoma in a curling motion around the storm system to the north. The low-pressure center at the surface lay over central Kansas with pressure at 28.98 in (98.1 kPa). Well to the east, toward the Mississippi River, squall lines with attendant turbulence were in the process of forming. One squall line raced through the Ohio Valley, a second formed about noon on a line extending from Indiana to Tennessee, and a third would soon appear over Illinois and Missouri. The massive outbreak of tornadoes to follow was associated with the squall lines.

The first reported tornado touched down at 9:30 A.M. near Lebanon, Indiana, a fateful omen of what was in store for the states between the Mississippi River and the crest of the Appalachians that day and evening. Not until the period of maximum solar heating and resulting atmospheric instability in midafternoon, when the dry air from the Southwest was entrained into the circulation at high altitudes and the moist air from the South was drawn far northward along the surface of the earth, did the grand outbreak of severe turbulence and tornado formation take place.

For six hours thereafter, approximately twenty reports of severe weather occurrences per hour were received and seventy-five percent told of actual tornadoes sweeping the surface of the ground. All three squall lines were producing tornadoes simultaneously. The most damaging and powerful struck at these locations: Xenia, Ohio (34 killed), at 3:40 P.M.; Brandenburg, Kentucky (31 killed), at 4:10 P.M.; Monticello, Indiana (19 killed), at 4:50 P.M.; and Guin, Tennessee (27 killed), at 9:04 P.M.

The greatest death toll from a single tornado track occurred between 6:30 and 8:45 P.M. in northwest Alabama and south-central Tennessee, where 63 people were killed. The longest track of the entire outbreak cut diagonally across northern Indiana for 134 mi (216 km), striking Monticello at 5:15 P.M. and killing 19.

By midnight a total of 125 tornadoes had struck in 11 states and caused 303 deaths and 5400 injuries. The pace slackened during the early morning hours of April 4, though 19 more tornadic storms occurred in Tennessee, North Carolina, West Virginia, and Virginia, resulting in four more deaths and about 100 injuries. The last tornadic activity occurred south of Lenoir, North Carolina, about 9:00 A.M., putting an end to 24 hours of tragedy from the sky.

Inside a Texas Tornado

Captain Roy S. Hall, U.S. Army, retired, had devoted much of his time to a study of meteorology as a hobby. When a massive tornado funnel struck his home on May 3, 1948, he was able to describe what he saw and experienced with the eye of a trained observer. He lived in the community of McKinney, Collins County, located about 30 mi (48 km) north of downtown Dallas in northeast Texas. The tornado did more than $2 million damage to over 300 homes and buildings. Captain Hall's full account appeared in the June 1951 issue of *Weatherwise* magazine.

The tornado moving from the west passed just to the north, placing the Hall home on the southern edge of the funnel. As it approached, Captain Hall described the "vivid lightning and rending crashes," but upon its arrival there was "a decided lull in the screeching roar outside." Hall described the next few seconds:

> In another minute the low cloud passed close overhead, and the dusk of early evening enveloped us. I turned to go in, and as I went up the porch steps hailstones the size of tennis balls began falling on the house and in the yard. These made my heart sink, for they almost invariably fall in the forefront of a tornado. They came down sparsely, one on about each square yard, but they made a most hideous bang and clatter, and I knew some of them were going all the way through our shingled roof. We all went into the west bedroom.
>
> Lightning was striking all around the house now, adding its horror to the fast-rising din. As my wife snapped on the overhead light, a gust of wind and rain hit the west wall of the room with a crash. My wife was pointing to the west wall. "The wall's blown in!" She had to scream to make herself heard. I could see that it had slipped inward six inches or more at the ceiling, and was vibrating under the wind pressure. Drops of water were hitting my face across the room. I tried to assure her. "That gust always comes ahead of a rainsquall," I shouted.
>
> But there was no abatement in the deafening hubbub outside. I knew it was growing in intensity by the second, and realized that a tornado was right on us. I yelled in my wife's ear:
>
> "Everybody in the back room! Get under the bed!"

* * *

And then very suddenly, when I was in the middle of the room, there was no noise of any kind. It had ceased exactly as if hands had been placed over my ears, cutting off all sound, except for the extraordinary hard pulse beats in my ears and head, a sensation I had never experienced before in my life. But I could still feel the house tremble and shake under the impact of the wind. A little confused, I started over to look out the north door, when I saw it was growing lighter in the room.

The light, though, was so unnatural in appearance that I held the thought for a moment that the house was on fire. The illumination had a peculiar bluish tinge, but I could see plainly. I saw the window curtains lying flat against the ceiling, and saw loose papers and magazines packed in a big wad over the front door. Others were circling about the room, some on the floor and others off it. I came out of my bewilderment enough to make a break for the back of the house.

But I never made it. There came a tremendous jar, the floor slid viciously under my feet, and I was almost thrown down. My hat, which I had not removed, was yanked off my head, and all around objects flashed upward. I sensed that the roof of the house was gone.

As I gained footing another jarring wham caught me, and I found myself on my back over in the fireplace, and the west wall of the room right down on top of me. The "whams" were just that. Instead of being blown inward with a rending crash of timbers, as one would expect of a cyclonic wind, the side of the room came in as if driven by one mighty blow of a gigantic sledge hammer. One moment the wall stood. The next it had been demolished. The destruction had been so instantaneous that I retained no memory of its progress. I was standing, and then I was down, 10 feet away. What happened between, I failed to grasp or to sense.

By a quirk of fate I was not seriously injured, and as soon as I had my senses about me I clawed up through the wreckage, and crawled around and through the hole where the east door had been. I could tell by the bluish-white light that the roof and ceiling of this room were gone also. I almost ran over my four-year-old daughter, who was coming to see about me. Grabbing her up, I was instantly thrown down on my side by a quick side-shift of the floor. I placed her face down, and leaned above her as a protection against flying debris and falling walls.

I knew the house had been lifted from its foundation, and feared it was being carried through the air. Sitting, facing southward, I saw the wall of the room bulge outward and go down. I saw it go, and felt the shock, but still there was no sound. Somehow,

I could not collect my senses enough to crawl to the small, stout back room, six feet away, and sat waiting for another of those pile-driver blasts to sweep the rest of the house away.

After a moment or so of this, I became conscious that I was looking at my neighbor's house, standing unharmed 100 feet to the south. Beyond I could see others, apparently intact. But above all this, I sensed a vast relief when I saw that we were still on on the ground. The house had been jammed back against trees on the east and south and had stopped, partly off its foundation.

The period of relief I experienced, however, was a very short one. Sixty feet south of our house something had billowed down from above, and stood fairly motionless, save a slow up-and-down pulsation. It presented a curved face, with the *concave* part toward me, with a bottom rim that was almost level, and was not moving either toward or away from our house. I was too dumbfounded for a second, even to try to fathom its nature, and then it burst on my rather befuddled brain with a paralyzing shock. It was the lower end of the tornado funnel! I was looking at its inside, and we were, at the moment, within the tornado itself!

The bottom of the rim was about 20 feet off the ground, and had doubtless a few moments before destroyed our house as it passed. The interior of the funnel was hollow; the rim itself appearing to be not over 10 feet in thickness and, owing possibly to the light within the funnel, appeared perfectly opaque. Its inside was so slick and even that it resembled the interior of a glazed standpipe. The rim had another motion which I was, for a moment, too dazzled to grasp. Presently I did. The whole thing was rotating, shooting past from right to left with incredible velocity.

I lay back on my left elbow, to afford the baby better protection, and looked up. It is possible that in that upward glance my stricken eyes beheld something few have ever seen before and lived to tell about. I was looking far up the interior of a great tornado funnel! It extended upward for over a thousand feet, and was swaying gently, and bending slowly toward the southeast. Down at the bottom, judging from the circle in front of me, the funnel was about 150 yards across. Higher up it was larger, and seemed to be partly filled with a bright cloud, which shimmered like a fluorescent light. This brilliant cloud was in the middle of the funnel, not touching the sides, as I recall having seen the walls extending on up outside the cloud.

Up there too, where I could observe both the front and back of the funnel, the terrific whirling could be plainly seen. As the upper portion of the huge pipe swayed over, another phenomenon took place. It looked as if the whole column were composed of rings or layers, and when a higher ring moved on toward the southeast, the ring immediately below slipped over to get back under it. This rippling motion continued on down toward the lower tip.

If there was any debris in the wall of the funnel it was whirling so fast I could not see it. And if there was a vacuum inside the funnel, as is commonly believed, I was not aware of it. I do not recall having any difficulty in breathing, nor did I see any debris rushing up under the rim of the tornado, as there surely would have been had there been a vacuum. I am positive that the shell of the twister was not composed of wreckage, dirt or other debris. Air, it must have been, thrown out into a hollow tube by centrifugal force. But if this is true, why was there no vacuum, and why was the wall opaque?

When the wave-like motion reached the lower tip, the far edge of the funnel was forced downward and jerked toward the southeast. This edge, in passing, touched the roof of my neighbor's house and flicked the building away like a flash of light. Where, an instant before, had stood a recently constructed home, now remained one small room with no roof. The house, as a whole, did not resist the tornado for the fractional part of a second. When the funnel touched it, the building dissolved, the various parts shooting off to the left like sparks from an emery wheel.

Line of thunderstorms approaching radar site from northwest. Enhanced images enable observer to judge intensity of the rainfall. ESSA photo.

June

Tell you what I like the best —
'Long about knee-deep in June,
'Bout the time strawberries melts
On the vine, — some afternoon
Like to jes' git out and rest,
And not work at nothin' else!
— James Whitcomb Riley, "Knee-Deep in June"

Calm weather in June
Sets corn in tune.

If June be sunny, harvest comes early.

A cold and wet June spoils the rest of the year.

If north wind blows in June, good rye harvest.

If St. Vitus' Day [June 15] be rainy weather,
It will rain for thirty days together.
— All from Richard Inwards,
Weather Lore (1898)

"June is bustin' out all over, / All over the meadow and the hill," runs the boisterous chorus of the famous Rodgers and Hammerstein song from *Carousel*. The reason for Nature's burst of life lies in the elevated angle of the rays of the sun, now at their northernmost position, where the celestial equator intersects the Tropic of Cancer at 23°27′ on or about June 21. The sun is then said to stand still at the solstice point (from the Latin *soltitium*: *sol* or sun, and *sistere* or to stand still). The direct overhead rays of the sun then fall on the parallel passing over southern Baja California, across Mexico just north of a Mazatlan–Tampico line, and through the Florida Straits. Key West stands just one degree north of this line; thus, no point in the 48 conterminous states ever receives the rays of the sun from the zenith point at 90° overhead.

The heat influence of the high angle of the sun and the greater duration of sunlight have important weather effects everywhere in North America. The last remaining snow melts from the tundras of Canada, so there are no more cold source regions for conditioning frigid polar air. The land becomes warmer in relation to the oceans, causing cooling sea breezes and sometimes fog along the seacoasts. As the interior is heated and the surface air rises, an inward flow of airstreams from the ocean takes place. Low-pressure cyclones tend to form over the land, while anticyclones reign over the ocean waters. The American version of the monsoon is in operation.

The Pacific High spreads laterally across the ocean along the latitude of San Diego. Its northeast arm extends over the coasts of Washington and British Columbia, causing a fair-weather regime there. The increased pressure gradient between the arm of the Pacific High and the now well-developed thermal low over the Southwest results in a strong northwest flow along coastal California, creating the comfortably cool and dry summer climate of that region. The Aleutian Low, beginning its annual westward migration toward the Siberian mainland, is centered in the Bering Sea, so the mainland of Alaska and western Canada enjoy a less stormy period.

In the Atlantic Ocean, the main axis of the Azores–Bermuda High lies along the latitude of Jacksonville, Florida, and while increasing in strength encroaches north toward the erstwhile haunts of the Iceland–Greenland Low. The latter shifts west also, into the waters between southern Greenland and northern Labrador, and has diminished greatly as storm generator for the North Atlantic.

June storm activity still retains many of the characteristics of spring in its distribution. The main storm track across the Great Lakes and down the St. Lawrence Valley remains virtually unchanged, as it has since January. Contrary to conditions prevailing in winter, the Intermountain region now spawns frequent storm systems, as do the central Great Plains. The Alberta track lies farther north than in early spring and becomes quite active, now favoring an eastward movement across Hudson Bay, where a branch splits northeast toward the major center of cyclonic action in the Iceland–Greenland Low over Davis Strait. In some years the beginning of the tropical storm season comes in June, when centers develop in the western Caribbean Sea and in the Gulf of Mexico, though several years may pass without the formation of a significant tropical storm so early in the season.

The main route of anticyclones across the country in June runs from west to east between latitudes 40°N and 50°N. They originate in the north Pacific High or its extension over British Columbia, move across the northern Rockies and northern Plains, and pass directly over the Great Lakes. They are joined there by an occasional polar anticyclone moving down from west-central Canada. The combined track continues east to reach the Atlantic Ocean over southern New England. These are fair-weather systems and are responsible for the delightful spells of fine weather that often occur in June before the heat and humidity of high summer arrive.

The influence of thundershowers in enhancing the rainfall of the midcontinent is clearly shown by the June precipitation distribution map. A large section of the country from the central Appalachians westward to the 100°W meridian and from Lake Superior to Arkansas are now enclosed by the 4.0-in (102-mm) rainfall line. Another such area extends along the coastline from Cape Hatteras to Florida and around the Gulf Coast to northeast Texas, though the interiors of the Gulf States generally have less than 4.0 in (102 mm). The commencement of the tropical rainy season is evident in southwest Florida, where the expected precipitation rises above 8.0 in (203 mm) in June. West of the Rocky Mountains there is less rain in June than in the preceding spring months, except in the interior of Washington and Oregon where June gives ample moisture before the annual summer dropoff. In the Southwest desert areas June is one of the driest months of the year.

Summer temperatures have arrived for most of the country by June. The advance of the mean 70°F (21°C) isotherm now runs from New York City to the Lower Great Lakes and on to northeast Nebraska, where it turns southwest to southern New Mexico. The 80°F (27°C) area now covers most of Florida, the central Gulf Coast, and much of Texas. Sizable portions of the valleys of the Southwest are enclosed by the 80°F (27°C) isotherm. The Pacific states enjoy their summer regime: comfortably cool on the coast and blazing hot in the interior.

Temperature extremes in June: maximum 127°F (53°C) at Fort Mohave, Arizona, on June 15, 1896; minimum 2°F (−17°C) at Tamarack, California, on June 13, 1907.

June is an important month in corn development and general farm progress. Cotton usually is blooming and fruiting in southern and central parts of the Cotton Belt and forming squares farther north; corn is silking and tasseling as far north as Virginia and Kansas by the close of the month. Winter grains are maturing north to central Pennsylvania and Nebraska and heading and blooming in the northern states. The planting of corn is often completed in the latter districts early in the month, and the cultivation of late potatoes and truck crops is under way in the North. The harvest of rye and barley advances except in the northern Plains.

June

1 1812 New Haven had latest blossoming of apple trees in period from 1794 to present, an effect of the "Cold Years 1812–1817."

2 1889 Great flood on Potomac at Washington, D.C.; took out span of Long Bridge; flooded streets near river; stage not equaled until March 1936.

3 1860 Great Comanche Tornado; commenced near Cedar Rapids, Iowa, ended over Lake Michigan; 175 killed; destroyed town of Comanche, Iowa, on Mississippi River.

4 1825 Early-season hurricane from Cuba to New England; great shipping losses on coast; damage at Charleston, S.C., on June 3, at New York City on the fourth.

5 1859 Great June frost from Iowa to New England; 25°F (−4°C) in New York State; 2-in (5-cm) snow in Ohio; wheat and buckwheat damaged.

6 1816 After 90°F (32°C) weather on June 5, cold front dropped temperature 49F (27C) degrees at Salem, Mass., to start "Year Without a Summer."

7 1816 Famous June snow; Danville, Vt., had drifts of snow and sleet 20 in (50 cm) deep; highlands white all day; flurries as far south as Salem and Boston area.

8 1966 Topeka Tornado in Kansas struck downtown area near capitol; 17 deaths; $100 million damage.

9 1953 Worcester County Tornado in Massachusetts; East's most deadly, with 90 victims; property damage placed at $53 million, then a record high.

10 1752 Traditional date for Benjamin Franklin's narrowly missing electrocution when flying kite in thunderstorm to determine whether lightning was related to electricity.

11 1842 Late-season snowstorm in northern New England; 11 in (28 cm) reported at Berlin, N.H., 10 in (25 cm) at Irasburg, Vt.

12 1884 Greatest June rainstorm ever at Los Angeles; 0.87 in (22 mm) on June 11–12; snow and hail fell at Pasadena on June 14.

13 1907 Temperature dropped to 2°F (−17°C) at Tamarack, Calif., lowest ever in U.S. in June.

14 1903 Heppner Disaster caused by cloudburst in hills that sent flood down Willow Creek in north Oregon; one-third of town swept away; 236 killed; $100 million damage.

15 1662 Drought at Salem, Mass.; fast day held with prayers for rain — "The Lord gave a speedy answer."

16 1806 Greatest of American total eclipses; path from southern California to Massachusetts; perfect weather in New England for viewing five-minute spectacle.

17 1859 Famous Santa Ana Wind at Santa Barbara; "grew very hot after 2:00 P.M.; fine dust or pulverized clay filled air; fruit roasted on one side"; temperature said to be 133°F (56°C).

1882 Grinnell Tornado struck Iowa college town; skip track covered 200 mi (322 km); 60 dead; $1 million damage.

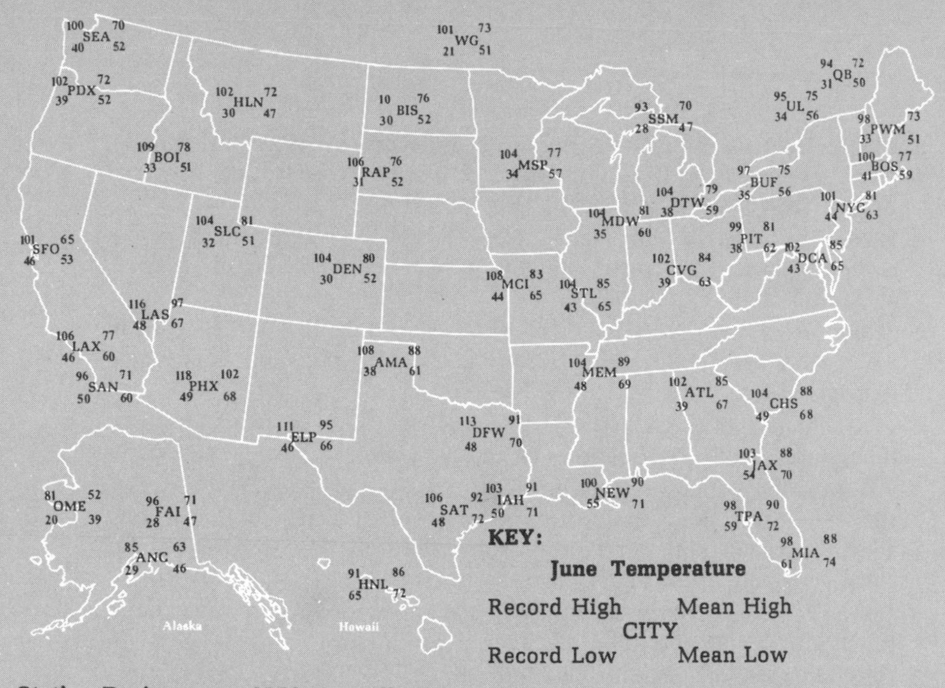

Station Designators: AMA Amarillo TX; **ANC** Anchorage AK; **ATL** Atlanta GA; **BIS** Bismarck ND; **BOI** Boise ID; **BOS** Boston MA; **BUF** Buffalo NY; **CHS** Charleston SC; **CVG** Cincinnati OH; **DCA** Washington DC; **DEN** Denver CO; **DFW** Dallas-Fort Worth TX; **DTW** Detroit MI; **ELP** El Paso TX; **FAI** Fairbanks AK; **HLN** Helena MT; **HNL** Honolulu HI; **IAH** Houston TX; **JAX** Jacksonville FL; **LAS** Las Vegas NV; **LAX** Los Angeles CA; **MCI** Kansas City MO; **MDW** Chicago IL; **MEM** Memphis TN; **MIA** Miami FL; **MSP** Minneapolis-St. Paul MN; **NEW** New Orleans LA; **NYC** New York NY; **OME** Nome AK; **PDX** Portland OR; **PHX** Phoenix AZ; **PIT** Pittsburgh PA; **PWN** Portland ME; **QB** Quebec QUE; **RAP** Rapid City SD; **SAN** San Diego CA; **SAT** San Antonio TX; **SEA** Seattle WA; **SFO** San Francisco CA; **SLC** Salt Lake City UT; **SSM** Sault Ste. Marie MI; **STL** St. Louis MO; **TPA** Tampa FL; **UL** Montreal QUE; **WG** Winnipeg MAN; **YC** Calgary ALB.

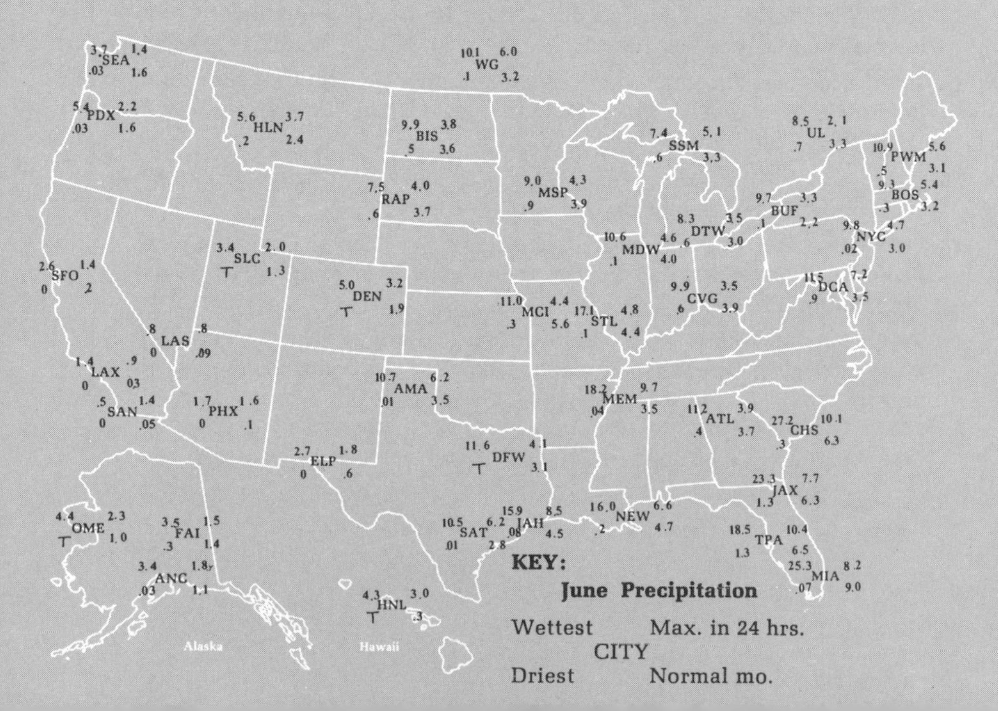

18 1875 Severe coastal storm, possible hurricane, from Cape Cod to Nova Scotia; Eastport, Maine, reported 57 mi/h (92 km/h) winds.

19 1835 Tornado tore through center of New Brunswick, N.J.; debris fell on Manhattan Island; 5 killed; studied on spot by William Redfield and James Espy, meteorologists.

20 1862 Civil War drought in South at its height: "The failure of oats in this region is total. Some wheat will be made but the crop is light and inferior" — Starksville, Ga.

21 1886 Destructive hurricane hit Apalachicola–Tallahassee area on summer solstice; extensive damage in Florida and throughout the Southeast.

1954 Severe hailstorm at Wichita and vicinity; $9 million damage.

1954 Floods on Iowa rivers; 10-in (250-mm) rain in four days; Des Moines exceeded previous record high stage by 3 ft (0.9 m).

22 1972 Most costly weather disaster in U.S. to that date: ex-hurricane Agnes loosed deluge of 12 in (305 mm) over Pennsylvania and New York; Wilkes-Barre on Susquehanna the principal sufferer; $2.1 billion loss; 122 victims.

23 1944 Tornado swarm, Ohio to Maryland, killed 150, mainly in West Virginia.

24 1816 Three days with 90°F (32°C) or more in Massachusetts after early June cold during "Year Without a Summer"; 99°F (37°C) at Salem, Mass.

25 1749 "A general fast on ye account of ye drought," in Massachusetts during famous dry spring when fields and villages burned.

26 1888 New York City's record heat wave of 14 consecutive days with daily average of 80°F (27°C) or more ended.

27 1901 Rain of fish from sky at Tiller's Ferry, S.C.; "hundreds of little fish swimming between cotton rows" after heavy shower — *Monthly Weather Review.*

28 1778 Battle of Monmouth in central New Jersey "fought in sweltering heat — the thermometer stood at 96° in the shade"; more casualties from heat than from bullets.

1924 Massive tornado hit Sandusky, Ohio, then swept across Lake Erie to smash Lorain; 85 killed; $12 million damage.

1980 Wichita Falls, Tex., set new record maximum of 117°F (47°C); old mark 113°F (45°C).

29 1954 Hurricane Alice dumped up to 27 in (686 mm) rain on Lower Rio Grande Valley; river at Laredo 12.6 ft (3.8 m) above previous highest; roadway of U.S. 90 bridge was 30 ft (9.1 m) below high water.

30 1886 Second hurricane in nine days hit Apalachicola–Tallahassee area; several killed, but damage light.

I wield the flail of the lashing hail,
And whiten the green plains under,
And then again I dissolve it in rain,
And laugh as I pass in thunder.

— Percy Bysshe Shelley, "The Cloud"

Thunderstorms

Thunderstorms provide an important portion of our country's water supply during the warm months of the year when precipitation from cyclonic activity is at a minimum. Where atmospheric conditions do not favor the formation of thunderstorms, such as the immediate Pacific Coast, drought conditions may prevail during the summer months. But along the Gulf and Atlantic coasts, the atmospheric structure is favorable for thunderstorm formation, and the resultant precipitation is normally adequate for growing crops without annual resort to irrigation. The agricultural prosperity of the great Corn Belt of the Midwest usually depends on precipitation from thunderstorms during the critical stages of the growing season.

A thunderstorm is a local storm cell invariably produced by a cumulonimbus cloud. It is always accompanied by lightning and thunder aloft and usually by strong gusts of wind, rain showers, and sometimes hail at the surface of the earth. In extreme cases, a funnel cloud or even a tornado may generate within the thunderstorm cloud. The activity of these weather elements in a single storm cell is usually of limited duration, often from ten to fifteen minutes, seldom over two hours.

Cause — The fundamental requirement for the development of a thunderstorm is an atmosphere that is unstable or potentially unstable if heat is added. Instability means that once a bubble of warm air is given an upward lift, it will become buoyant and continue to rise without additional impulse. The air bubble constantly finds itself in a colder environment and, like a hot-air balloon, is forced upward by the pressure of cooler, denser air surrounding it. This is the process of convection.

To quote Napier Shaw, "Thunderstorm is a gigantic, if comparatively slow, explosion of moist air, the latent heat of the moist air acting as fuel." Conditions favorable for convection arise from several circumstances. Daytime heating over flat land may result in rising thermal currents whose further ascent is stimulated by the unstable condition of the atmosphere. Air resting over rough or mountainous terrain may receive unequal heating and cause vertical circulations. Airstreams passing over rising high ground or a mountain range may be forced upward, orographically, to an unstable layer and continue the ascent to higher levels. Convergence of airstreams in a frontal zone during cyclonic activity may create vertical currents and increase instability.

Season — Thunderstorms have their maximum frequency during the warm months of the year, except along the Pacific Coast. Regions east of the Mississippi River experience their greatest number in July. Over much of the Great Plains region, June is the month of maximum frequency since the proper combination of moisture and heat is normally present. The northern Rocky Mountains and Intermountain regions have their maximum in July, while the southern Rockies favor August, when tropical Gulf air arrives. The interior of the Pacific Northwest reaches its high in June, before the Pacific High is firmly established over the area. The months from November to February have the greatest number of thunderstorms from central California northward along the coast to Washington, since frontal situations often lift the air over the coastal mountains and create the necessary unstable conditions. Southern California marks March and April as the months most favorable to thunderstorm development.

Area — Within the United States, there are two areas of maximum occurrence: (1) Locations from the crest of the Rocky Mountains in Colorado and New Mexico to the High Plains immediately to the east may expect 50 or more thunderstorms per year, and (2) the east Gulf Coast and the peninsula of Florida have a normal figure of 80 or more. Generally, the number decreases from south to north. The states bordering Canada normally witness fewer than 30, while the number drops to less than 20 along the New England coast. A low-frequency area stretches along the Pacific Coast. San Francisco Airport will report only one or two thunderstorms a year, though Blue Canyon, a mile high in the Sierra, has about twelve. Both Oregon and Washington report fewer than fifteen, the number increasing eastward over the region. Thunderstorms occur over the interior of Alaska, perhaps five times a year for a location such as Fairbanks, but very rarely along the seacoast, where winds from the cool ocean discourage the creation of convective currents.

Size — The storm may consist of only a single swelling cumulus cloud covering a few acres, emitting a spate of rain and one roll of thunder, or the storm structure may develop a supercell of great vertical height, reaching the top of the troposphere at approximately 60,000 ft (18,300 m). It may develop excessive rains of flash-flood potential, create straight-line blasts of damaging force, spawn a tornado or two, and cause heavy crop losses with destructive hail. A series of individual supercells may cover a land area of hundreds of square miles, or the cells may join in a line along a squall or cold front and extend laterally several hundred miles across the country, presenting a major hazard to aircraft.

Time of Day — The late afternoon and early evening hours, when heat is exerting its maximum effect and instability is present, have the greatest frequency of local convective thunderstorms. Frontal thunderstorms may occur at any time of day or night, though they, too, attain their greatest severity in the late afternoon or early evening when the day's heat adds to the available energy. Thunderstorms forming over the Rocky Mountains in daytime may drift eastward over the Great Plains during the night and still retain a high degree of severity past midnight, since their development is sustained by radiational cooling from the top of the cloud into the dry atmosphere above, maintaining instability within the cloud. In other areas, local thunderstorms usually decrease in intensity after sundown and dissipate by midnight.

Opposite: Cumulonimbus thunderhead spreading its anvil over mountains near Tucson, Arizona. University of Arizona photo courtesy of Dr. Louis Battan.

Formation — An unstable atmosphere encourages the continuation of convection. Eventually, the air in the rising thermal current will cool sufficiently to reach its dew point or condensation level; then its water vapor will pass from an invisible gaseous state into solid water droplets appearing as cloud. In most cases this is the end result of the process — a cumulus cloud. But if the rising thermal currents have sufficient energy, they will continue to ascend and expand laterally and upward to become a swelling cumulus cloud. Then other forces come into play.

A young thunderstorm feeds internally on another source of energy. Upon the formation of the visible cloud, the change of phase from vapor to water droplets releases latent heat, which adds to the heat energy available to increase the momentum of the storm's vertical development. The rate of addition of new energy is directly related to the amount of gaseous water vapor being converted into liquid water droplets or solid ice crystals. The created rain, hail, or snow begins to acquire size and weight and eventually falls downward within the cloud when it has attained sufficient mass to overcome the ascending currents inside the cloud.

The full-growth stage of the thunderstorm is reached when the main updraft is joined by one or more downdrafts created by the falling precipitation. The descending currents are fed and strengthened, as was the updraft, by the addition of entrained air and by evaporational cooling. Lightning accompanies the formation and descent of the precipitation, indicating a close relationship with formation and splitting of ice crystals and raindrops. At maturity, the thunderstorm cloud may measure several miles across in all directions at its base and may tower to altitudes of 40,000 to 60,000 ft (12,200 to 18,300 m). Some reach to the base of the stratosphere, where the vertical development halts and swift winds in the upper atmosphere tear away the cloud top and carry its shreds downwind many miles ahead of the advancing cumulonimbus, producing the familiar anvil-shaped cloud, the trademark of a mature cumulonimbus cloud. The storm is now at its most violent phase. Lightning and thunder are incessant aloft; melting snow, hail, and rain cascade downward; violent vertical currents prevail; and possibly a tornado funnel lowers toward the surface of the earth.

Decay — The storm cell soon passes its most violent stage and starts to break up. The cold downdrafts, upon reaching the surface, eventually strangle the warm updrafts that initiated the storm's development. With the rising air cut off, the cell is deprived of its source of sustenance. Precipitation gradually slackens off, and the violent downdrafts subside. The top of the thunderstorm cloud begins to disintegrate by spreading out laterally and moving downwind. Soon a more stable equilibrium with the surrounding atmosphere is attained.

Lightning — The atmosphere's most dazzling production is lightning, as thunder is its noisiest. The two have a close cause-and-effect relationship. Lightning may be described as an enormous spark, an electrical discharge of great potential. It originates in towering clouds of the cumulonimbus variety, which often develop into local storm factories as a result of the strong ascending and descending currents existing within the cloud. The splitting and collision of water droplets and ice crystals in these currents create opposite electrical charges that concentrate in different sections of the cloud, creating a giant storage battery. The upper regions of the cloud carry a net positive charge while the lower regions are negatively charged. Lightning is a discharge between the two, or between the lower portion of the cloud and the positively charged earth. Discharges from one cloud to another can also take place. Most lightning activity occurs within the cloud; it is estimated that only about 20 percent of the discharges are of the cloud-to-ground variety, so spectacular to earthlings.

The thunder, That deep and dreadful organ-pipe.
— William Shakespeare, *The Tempest,* III, iii, 97

And hark to the crashing, long and loud,
Of the chariot of God, in the thunder-cloud!
— William Cullen Bryant, *The Hurricane*

The heavens thundered and the air shone with frequent fire; and all things threatened men with instant death.
— Virgil, *Aeneid,* Book I, 1, 90

Thunder — The sound emitted by rapidly expanding gases along the channel of a lightning discharge is called thunder, from the Latin *tonare* and the Old Norse god *Thōrr*. Much of the energy of a lightning discharge is expended in heating the atmospheric gases in and immediately around the luminous channel. In a few microseconds, it rises to a local temperature of the order of 10,000° Centigrade, with the result that violent pressure waves are sent out, followed by a succession of rarefactions and compressions induced by the inherent elasticity of the air. These sound waves are heard as thunder.

Thunder is seldom audible at points farther than about 15 mi (24 km) from the lightning discharge, with 25 mi (40 km) being approximately the upper limit and 10 mi (16 km) a fairly typical range of audibility. At such distances, thunder has a characteristic rumbling sound of very low pitch. When heard at long distances, the pitch is low because of strong absorption and scattering of the high-frequency components of the original sound waves. The rumbling results chiefly from the varying arrival times of the sound waves emitted by the sinuous portions of the lightning channel, which are located at various distances from the observer, and, secondarily, from echoing and from the multiplicity of the strokes of a composite flash.

Since one's sighting of a lightning flash is almost instantaneous with its occurrence and the sound normally travels about 1090 ft (332 m) per second, it takes about five seconds for the thunder to travel a mile. One can get an approximation in miles of the distance of thunder by counting the number of seconds between flash and sound and dividing by five.

Inside a Carolina Thunderstorm

Marine pilot William H. Rankin was making a Visual Flight Rules trip from Massachusetts to South Carolina in 1960 when his F8U plane flamed out at 47,000 ft (14,326 m) over North Carolina. His emergency power package failed to respond. Realizing he was without radio, instruments, and power controls, and the stick was frozen in the neutral position, pilot Rankin decided to eject before descending in uncontrolled flight into the top of a thunderstorm. He ejected directly over an extensive thunderhead estimated to have been 100 mi (161 km) in diameter. The plane's speed at the time of ejection was approximately 210 knots (242 mi/h or 389 km/h).

His dramatic 40-minute, 9-mi (14-km) descent through the violent thunderstorm was given wide coverage in the nation's press. A more detailed account of the pilot's experiences after ejection was compiled from his statements during the accident investigation and published in *Approach,* the Naval Aviation Safety Review. The part describing his entrance into and descent through the thunderstorm follows:

> As the turbulence started, I was pelted all over by hail. Then I fell a little bit more and I seemed to be caught in a violent up-draft. I had the feeling that I was being tossed around ... that I was actually going around in a loop and I was looping over my canopy like being on the end of a centrifuge. I got sick in the turbulence and heaved.
>
> Sometimes I could see the canopy and sometimes I couldn't. The tossing and turbulence was so violent it is difficult to describe. I went up and down ... I was buffeted about in all directions ... at times I felt like I was going sideways. One time I hit a very rough blast of air — I went soaring back up and got in a very severe hailstorm. I remember the hail beating on my helmet, I had the feeling it would tear my canopy up. The next thing I knew I was in rain so heavy I felt like I was standing under a waterfall. I had my mask loose and the water was so great that when I tried to inhale I got water with the air like I was swimming. It seems to me that sometime in the storm I noticed my watch and was surprised that it had stayed with me. I'm not sure but I think I was able to tell the time by the luminous dial ... I believe it was around 1815.
>
> At one time during an up or down draft, the parachute canopy collapsed and came down over me like a big sheet. I could see my legs in the shroudlines. This gave me some concern — I thought maybe the chute wouldn't blossom again properly and since the hail seemed to be larger now I was afraid it might damage the canopy and put holes in it. I fell and the canopy blossomed again. I felt the risers and everything seemed all right.
>
> At this time, I looked down and saw what appeared to be a big black elevator shaft. Then I felt like I had been hit by a blast of compressed air and I went soaring back up again — up and down — sideways. How much of this soaring went on I don't know. I had the feeling that if it went on much longer I was not going

to maintain consciousness. I was being tossed around and beaten around and I wasn't quite sure how much more I could take.

The violence was so great that I thought that if it doesn't stop soon, my gear will come apart . . . my chute will come apart . . . and my straps will break . . . I will come apart. Stretching . . . twisting . . . slamming . . . the turbulence of this thunderstorm was so violent I have nothing to compare it with. I became quite airsick and I had considerable vertigo. Again I had the feeling that I couldn't take much more of this but if I could only hold out a little while longer, I would be falling out of the roughest part of the storm.

The lightning was so severe that I kept my eyes closed most of the time. Even with my eyelids closed, there was a blinding reddish-white light when the lightning flashed. I felt rather than heard the thunder; it just about burst my eardrums. As I recall, I had the feeling that I was in the upper part of the storm because the lightning seemed to be just flashes. As I descended, I seemed to see big streaks headed toward the earth. All of a sudden I realized that it was getting a little calmer and I was probably descending below the storm. The turbulence grew less, then ceased and I realized I was below the storm. The rain continued, the air was smooth and I started thinking about my landing.

Lightning Bolts — The lightning bolt causes more direct deaths in the United States than any other stormy weather phenomenon. Lightning casualties do not receive the widespread publicity given to tornado or hurricane fatalities because all but a few lightning deaths are single events, when one person is killed by a single lightning strike.

Between January 1940 and December 1976, lightning (either directly or indirectly) was reported to have killed more than 7500 Americans and injured more than 20,000 others, giving annual averages of approximately 200 killed and 550 injured. Incomplete reporting has resulted in conservative numbers, and it is estimated that the annual average death toll from lightning is probably double that number.

Another survey concluded that from 75 to 85 percent of those killed annually are males, since their business and entertainment activities take them out of doors regularly. About 22 percent of the victims are children under 18 years of age. About 70 percent of all injuries and fatalities occur in the afternoon, when thunderstorms are most numerous, and the season of greatest frequency is from May through August.

The occurrences are nationwide, with the minimum in the three Pacific Coast states and a maximum in the Gulf States. Only one state, Florida, appears to have an annual average of lightning deaths in excess of 10. Seven states — Florida, Louisiana, Michigan, Mississippi, New York, North Carolina, and Pennsylvania — have an average of 10 or more lightning injuries each year.

Fifty incidents resulted in six or more casualties each during a nine-year period. Twelve occurred at military bases and resulted in 182 injuries and 5 deaths. Five incidents involving persons at work resulted in 32 injuries and 3 deaths. The majority of group casualties with six or more occurred while the victims were participating in some form of recreational activity. A total of 426 injuries and 19 deaths occurred during such events as camping trips, football and baseball games, fairs, and golf matches.

The favorite lightning target is a tree, and the greatest single cause of forest fires in the western United States is lightning. Each year about 7500 forest fires are started by lightning, causing a loss of about $25 million annually. Nearly two million acres (0.8 million ha) of valuable timber land — an area larger than the entire state of Delaware — went up in smoke from 1931 to 1953 because of lightning.

Lightning fires are seasonal, varying with thunderstorm activity. The winter months are quiet, but lightning reaches a maximum in late July and August, coordinating with the peak of the annual dry season in the Northwest. Lightning forest fires also vary from year to year. For instance, in the northern Rockies in 1943 only 538 fires were set by lightning, but in 1940 over 3000 fires were started in the same area, and in one hectic month, July 1940, nearly 2000 fires were ignited. The big day came on July 12, with no fewer than 335 fires.

In the western mountains lightning strikes are more frequent at higher elevations. The favored spot in the United States is the Clearwater Forest in Idaho just above 7000 ft (2134 m), where the frequency is 12 times that for all other western forests.

Most Costly Lightning Stroke — A single bolt struck the former Naval Ammunition Depot on the grounds of the present Picatinny Army Arsenal at Lake Denmark, Morris County, in northern New Jersey on July 10, 1926, triggering a series of explosions that ended in a major catastrophe. A red ball of fire leaped into the air. All buildings within 2700 ft (823 m) were destroyed and 16 people died. Debris fell as far as 22 mi (35 km) away. Property damage in the amount of $70 million resulted from the single lightning stroke.

Some Unusual Lightning Events

1900, July 5 — Lightning struck the Standard Oil refinery at Bayonne, New Jersey, on Newark Bay. A spectacular three-day fire resulted, with burning oil floating on the bay waters. The smoke column rose to 13,000 ft (3962 m). Many whirlwinds formed. One person was killed and three others seriously injured by exploding tanks. Damage was $2 million.

1917, August 1 — After a very dry spring and summer, an electrical storm raged over Trinity County, California. Little or no rain fell and 80 forest fires were started. Lightning was reported to have struck 150 times in an area of about five sq mi (13 km/sq). A fire warden said that the country "looked like a vast Christmas tree as various trees blazed into light after being struck."

1918, July 22 — A flock of 504 sheep was killed by one bolt in the Wasatch National Park in Utah.

1926, April 7 — Lightning at San Luis Obispo, California, started a disastrous oil fire that lasted five days, spread over 900 acres (364 ha), and burned nearly six million barrels of oil. Two persons were killed. Property losses were $15 million. Hundreds of small whirlwinds formed over the inferno.

1932, April 22 — Fifty-two large wild geese were killed by a lightning flash during an electrical storm. It was seen to strike into a large flock of birds while they were passing over Elgin, Manitoba, and 52 tumbled to earth. They were picked up and distributed to townspeople for goose dinners.

1937, May 6 — German zeppelin *Hindenburg* was destroyed at Lakehurst, New Jersey, with thunderstorm in vicinity. Residual static electricity may have ignited a leaking hydrogen cell when landing lines made contact with the ground. Sabotage of the Nazi ship was also thought to have been a possible cause. No definitive solution of the cause was agreed upon by the boards of inquiry, nor by subsequent private investigations. Thirty-six were killed.

1937, September 4 — Four golfers were killed and three felled by lightning stroke at Long Vue Country Club at Pittsburgh, Pennsylvania.

1941, July 12 — Three caddies were killed by lightning on golf course at Louisville, Kentucky.

1943, August 10 — Lightning killed six soldiers on drill field at Fort Belvoir, Virginia.

1949, July 31 — Lightning struck baseball diamond at Baker, Florida, while game was in progress. The bolt dug a ditch 20 ft [6 m] long through the infield, killed the shortstop and third baseman, fatally injured the second baseman, and injured 50 people in a crowd of 300 spectators.

1961, July 12 — Nine tobacco workers, aged 13 to 70, were struck while taking shelter in a tobacco barn 3 mi (4.8 km) south of Clinton in Sampson County, North Carolina. Lightning apparently struck metal heating system against which victims were leaning. Only one survived. Three other workers under open shelter outside the building were unhurt.

1963, December 8 — Jet liner exploded near Elkton, Maryland, killing all 81 aboard. Lightning is believed to have caused the explosion of residual fuel vapor in one of the outboard wing tanks as plane passed through a vicious, out-of-season thunderstorm.

1967, August 15 — "The 56,000-acre Sundance fire in northern Idaho, started by lightning, fed by massive expanse of forest fuels made bone dry by one of the hottest, driest summers in recent memory, and driven by winds of more than 50 mph [80 km/h], traveled across the Selkirk Mountains like a giant tornado. It spread 20 mi [32 km] in 12 hours. Funnel-shaped vortices of flame, calculated to be whirling at speeds up to 300 mph [483 km/h], flung giant trees around like matchsticks. The research team estimated that peak energy release of 3–5-mi [5–8-km] flame front was equivalent to that of a 20-kiloton atomic bomb exploding every two minutes. Firebrands were carried as much as 10 mi [16 km] in advance of the flame front to ignite new fires." — J. S. Barrows, *Forest Fire Science*

1968, August 24 — Seventy-two persons were injured and two were killed in Crawford County, Pennsylvania, when lightning struck tent poles at the Crawford County Fairgrounds.

1970, September 7 — Lightning struck a school gridiron while game was in progress at St. Petersburg, Florida. Two players in huddle were killed and 21 others injured.

1971, July 18 — Four cyclists took shelter in a grove of trees and were killed by lightning in Taos County, New Mexico.

1971, September 6 — Four men stringing tobacco in a barn in Monroe County, Kentucky, were killed when lightning struck the metal roof. On the same day, several hours later, four more men were killed under the same circumstances in the nearby county of Macon in Tennessee.

1972, September 24 — A 15-year-old boy was struck by lightning while carrying 35 pieces of dynamite at a logging operation near Waldport, Oregon.

1974, July 17 — Norma Jean, a 6500-pound elephant, was killed by a lightning stroke at Oquawka, Illinois, while chained to a tree in the town square where a small circus had pitched its tent. The lightning struck the tree and passed through the chain.

1975, June 27 — Champion golfer Lee Trevino and two companions were struck by lightning at the Butler National Golf Course at Oak Brook, Illinois, while playing in the Western Open. Trevino was sitting under an umbrella and leaning on his golf bag when the bolt struck the bag. His back was slightly burned. All three were stunned and taken to the hospital for an overnight stay.

1975, August 23 — Ninety persons were injured when lightning struck a campground in Ingham County, Michigan, at 4:45 P.M.

1977, June 26 — Park Ranger Roy C. Sullivan of Virginia was struck by lightning for the seventh time, earning him the title of "the human lightning conductor." His attraction to lightning commenced in 1942 (lost big toenail), and then after a long lapse resumed in 1969 (lost eyebrows), in July 1970 (left shoulder seared), in April 1972 (hair set on fire), in August 1973 (new hair fired and legs seared), in June 1976 (ankle injured), and in June 1977 (sent to hospital with chest and stomach burns). We do not know what kind of metal object Sullivan was accustomed to carry with him.

1980, May to September — Lightning generated in the convective clouds following the eruption of Mt. St. Helens in Washington on May 18, 1980, started hundreds of fires in the snags and debris buried under volcanic ash. They burned until extinguished by autumnal rains.

Hailstorms

Precipitation in the form of solid spheres of ice or irregular frozen conglomerates contributes a spectacular feature to the atmospheric antics during the warmer portion of the year. Hail originates in a convective cloud of the cumulonimbus type, which develops in the strong updrafts attending frontal turbulence or a local thunderstorm. The most prolific hailstorms result from the formation of supercells in a thunderstorm whose enormous updrafts permit large hailstones to grow by accretion over periods of many minutes. Over the Great Plains, where hailstorms are frequent, most of the crop damage is caused by only a few supercell storms each season.

Most hailstorms cover only a limited area with a scattering of stones falling, but when fully developed they may loose vast quantities of icy matter over a considerable territory. Often hail falls over long areas from 0.5 to 2.0 mi (0.8 to 3.2 km) wide and from 5 to 10 mi (8 to 16 km) or more long as the developing storm cloud moves downwind. These areas, known as hailstreaks, can turn white from the accumulating hailstones and be covered to a considerable depth by an icy mass. Total destruction of crops may result, and injury and even death may come to humans and animals when caught in the open by the bombardment from the sky.

Hailstones — Hailstones exhibit a great variety of shapes and structures. A single stone may range from the size of a pea to that of a grapefruit, or from less than 0.25 to more than 5.0 in (0.6–

Gigantic hailstone as compared with hen's egg. NCAR photo.

12.7 cm) in diameter. They may be spherical, conical, or quite irregular in shape. Spherical stones, the most common form, exhibit a layered interior structure, sometimes resembling an onion, with layers composed of clear ice alternating with layers of rime ice that looks white by virtue of many tiny air bubbles.

Hailstones grow by accretion around a small ice pellet when in the presence of supercooled water vapor in the cloud. They must remain aloft long enough to acquire a size sufficient to withstand complete melting in their descent through warmer air. They will continue to grow in size as long as the strong updrafts in the cloud support them. Often hailstones are tossed out of the chimney-effect updraft into descending currents, or the supporting power of the updraft weakens; the ice then commences its descent as a potentially damaging missile.

When hailstones are carried up and down in vertical currents several times, they melt a little in each descent and acquire a new sheathing of ice during their ascent, producing stones with varying textures and appearances. Some develop protruding lobes that resemble feet, probably the result of a spinning action in their descent. Sometimes several hailstones freeze together, forming irregular chunks of ice that often smash into pieces upon impact with the earth. As is the case with their distant relatives, the snowflakes, hailstones can have a vast variety of shapes, structures, and appearances.

Largest Hailstone — On September 3, 1970, a hailstone measuring 17.5 in (45 cm) in circumference and weighing 1.671 lb (758 gm) was picked up and photographed at Coffeyville in southeast Kansas. The largest stone previously documented by the National Weather Service in the United States was almost the same size; found at Potter in southwestern Nebraska on July 6, 1928, it had a circumference of 17 in (43.2 cm) and weighed 1.51 lb (680 gm). The *Monthly Weather Review* reported a comparable stone falling at Dubuque in eastern Iowa on June 16, 1882, which also measured 17 in (43.2 cm) in circumference and weighed 1.75 lb (794 gm).

Accumulations — A severe storm at Selden in northwest Kansas on June 3, 1959, left an area 9 × 6 mi (15 × 10 km) covered with hailstones to a depth of 18 in (46 cm). The hail fell for 85 minutes and did $500,000 damage. Hailstones tend to be swept downhill by accompanying heavy rain and to accumulate in drifts. Piles 6 ft (1.8 m) high were reported by Iowa farm editor Henry Wallace at Orient on August 6, 1980, and, where protected from the sun, they remained on the ground for 26 days. A massive fall in Nodaway County, northwest Missouri, on September 5, 1898, left hail on the ground for a period of 52 days until October 27 when enough remained to make ice cream. The ice-clogged fields remained unworkable for two weeks after the storm. On some occasions snowplows have been called out to clear highways after a heavy hail fall on the High Plains.

Intensity — The "intensity" of hail is what produces damage. It is a function of the number of stones, their sizes, and the force of the wind driving them. Comparative studies of crop damage have demonstrated that hailstorms during the peak of the loss season in eastern Colorado are 18 times more intense than during the typical crop season storms in Illinois. Away from the western Great Plains, intensity decreases generally, though damage will vary according to the type of crops and the way they are planted.

Season — The time of year of maximum hail activity varies in different parts of the country. East of the Great Plains, maximum frequency comes in the spring months, starting in the Deep South in March and moving northward into the Midwest and Northeast in May. On the Great Plains, especially in the lee of the Rocky Mountains, the maximum hail activity occurs in the summer months, from June to August. The Great Lakes region is the only place in North America where maximum hail occurs in the autumn months. Along the West Coast, certain areas have their high hail season in late winter or spring. The season of hail occurrence is important in relation to stages of crop development.

Two exceptional out-of-season hailstorms hit the St. Louis area on December 11, 1949, and December 2, 1950; the latter covered the ground to a depth of 1.5 in (3.8 cm) and impeded traffic. In wintertime, small ice pellets (commonly called sleet) are often mistaken for hail; these are merely frozen raindrops, without layered structures.

Time of Day — The hours of hail occurrence have some interesting regional differences. In the states bordering the Mississippi River and eastward, the hours of maximum occurrence lie between 2:00 and 7:00 P.M., with a concentration between 2:00 and 5:00 P.M., the hours being much the same as for severe thunderstorms and tornadoes. On the High Plains in the lee of the mountains, the hours of maximum occurrence are from 12:00 noon to 3:00 P.M. and then during the afternoon and evening the storms move eastward across the region until reaching the Mississippi River, where there is a secondary period of maximum frequency from midnight to 3:00 A.M. Hail very seldom falls between 5:00 A.M. and 10:00 A.M. anywhere.

Hail Alley — The High Plains immediately east of the Rocky Mountains have the greatest frequency of hailstorms. This zone extends southeast from Alberta into Montana and then south to include adjacents parts of Wyoming, South Dakota, Colorado, Nebraska, Kansas, Oklahoma, Texas, and New Mexico. Cheyenne, lying to the east of the Laramie Range in Wyoming, averages nine to ten storms per season in its immediate vicinity. Hailstorms may occur anywhere in the United States when convective activity is present, moisture is available in the clouds, and the freezing level is relatively low. The immediate Pacific Coast has the least frequency, but the activity increases considerably in the interior mountains.

Length and Width of Hailstreaks — Hail areas in the larger storms are usually somewhat rectangular in shape, varying from 100 ft (30.5 m) to 0.5 to 2.0 mi (0.8–3.2 km) in width. An average hailstreak is about 0.5 mi (0.8 km) wide and about 5.0 mi (8.0 km) long. But hail swaths crossing many counties have been known to cover distances of 200 mi (322 km). These extensive paths are often composed of many individual hailstreaks, somewhat like the path of a skipping tornado.

Hailstorms are seldom well documented, since weather stations are generally widely separated, but the Illinois State Water Survey maintains a dense network of rain gauges and hailpads in central Illinois between Bloomington and Decatur. Fortunately, a very severe rain-, hail-, and windstorm passed over the array of instruments on May 15, 1968. One super hailstreak, with a maximum width of 19 mi (30.6 km) and a length of 51 mi (82 km), covered 788 sq mi (2049 sq km). At one moment the area of falling hail was 19 mi (31 km) wide and 10 mi (16 km) long and the hail was moving forward at 35 mi/h (56 km/h). Its entire duration was 90 minutes. The larger stones were 2.25 in (5.7 cm) in diameter, and the total production was 82 million cu ft (2.3 million cu m) of ice!

Annual Cost — Crop losses from hail average $773 million annually (in 1975 dollars), according to a recent survey. Property losses amount to $75 million. Fifty percent of all losses occur in the Great Plains from Texas to North Dakota. The intensity of hailfalls (hail and wind combined) is from five to fifteen times greater in this area than elsewhere in the United States. Since wheat is the principal crop in this region, it accounts for 51 percent of the total national loss. Other crops subject to frequent damage nationwide by percentage are: cotton 11, corn 10, soybeans 9, and tobacco 7. Approximately 25 percent of the losses are covered by hail insurance.

U.S. Most Fatal Hailstorm — The *South Carolina Gazette* carried the following dispatch in the issue of July 1, 1784:

> On the eighth of May last, a most extraordinary shower of hail, attended with thunder and lightning, fell in this district, and along the banks of the Wateree; the hail stones or rather pieces of ice, measured about nine inches in circumference; it killed several negroes, a great number of sheep, lambs, geese, and the feathered inhabitants of the woods without numbers: its greatest violence did not extend more than two miles in breadth, but where it began or ended is not known; within that space it stript trees of their leaves and even their bark, and every blade of grass was beat to the ground. But what is still more astonishing there are at this time [46 days later] many wagon loads of hailstones unmelted, lying in the hollows and gullies on the Wateree. The truth of the above facts can be testified by the inhabitants of the place.

Account of a Hailstorm at San Antonio, Texas

By Lieutenant George M. Bache, U.S.A.

> One of those terrific hail-storms often heard of, but seldom experienced, visited this city last evening [May 10, 1868]. Captain A., Doctor B., and myself were returning from a day's fishing some ten miles from San Antonio, in an army ambulance, about 7 P.M., when we first noticed indications of rain. Dark clouds were rising in the northwest, accompanied with a great deal of what we at first supposed to be heat lightning. This gradually became more vivid, the clouds blacker, the thunder began to make itself heard, and our first supposition of a light shower changed to a certain prospect of a severe storm.
>
> * * *
>
> Continuous blows on the head, body, and legs soon enabled us to realize the serious nature of our condition. Stones of ice of all

Opposite: Cornfield totally destroyed after hailstorm, July 10, 1975, in northwest Virginia. Leesburg Times Mirror photo.

shapes and of the size of the first, cut and bruised our bodies, and with our arms crossed above our heads we rushed to secure the slight protection of a mesquite bush, there being no trees on the prairie. We were each at different times knocked down by blows about the head; one of us, Captain A., three times. Cut, bleeding, bruised, and still with no prospect of abatement, not knowing how long such a phenomenon might last, nor how soon we might be rendered senseless, we felt our situation as by no means enviable.

* * *

In the mean time, the mules, which having again headed the storm in their fury were nearly stunned by repeated blows on the head and sides, came near us, being driven before the storm, but too much weakened to move rapidly. We took advantage of this and leaped into the ambulance, choosing the lesser evil. Providentially the wheels of the carriage became locked in a rail fence, and the mules were too much exhausted to do any more running. We put the seats over our heads, and thus protected, drenched, shivering with cold, and continually beaten on the legs and sides, we awaited the subsidence of the storm. The falling of the hail lasted twenty-two minutes, commencing at half past eight P.M. It was accompanied with heavy rain, bright blinding flashes of lightning, and a continuous roar, varied with sharp crashes, of thunder. The rain ceased with the hail, but fell very heavily again during the night, causing a rise of nearly twenty feet in the San Antonio river. The curtains of our ambulance were cut to ribbons, and we scarcely thought the mules would live through the tempest, but they did, and, though much bruised and stunned, brought us safely to town. With black eyes, bloody heads, smashed hats, bruised arms, and torn and muddy clothes, we appeared as if we had just come from a free fight and had been very badly used. Indeed, experience only could have convinced us that any one could have endured exposure to such a violent storm and lived.

* * *

A woman who, with her husband, had camped out on the prairie, had two ribs broken, and was thought to be fatally injured. A dog was killed outright; and there are numerous cases of cuts and bruises, more or less severe. There are also reports of fatal casualties to human life in the vicinity, which have not yet been authenticated. The momentum of the hail-stones is shown by the fact, as witnessed by myself, of a hole about four inches in diameter through both sides of a sheet-iron stovepipe which rose from the roof of a small out-house in the garden and did service as a chimney. Boards of fences were knocked off and split in pieces, and trees barked as if by cannon balls. I think the storm of hail not to have extended over a path of more than two miles in width. We hear of the storm having visited other places, but having no communication, save by stages and a semi-weekly paper, we have not yet learned its course. The hail came down at an angle of about 30° from the horizontal, and lowered the temperature from 90° to 64° Fahrenheit. The day had been close and sultry. The temperature again rose after the storm, and the stones on the ground were soon melted.

Individual North American Hail Fatalities — The first instance in National Weather Service records of a hail death occurred on May 13, 1939, near Lubbock, Texas, when a 39-year-old farmer died of injuries after being caught in the open during a severe hailstorm. The country's press have reported other hail fatalities: at Amwell, New Jersey, on June 10, 1742; in St. Charles County, Missouri, on September 11, 1863; in Broome, Quebec, on July 28, 1879; near Uvalde, Texas, on May 17, 1909; at Windsor, North Carolina, on May 10, 1931, and near Toronto, Ontario, on June 13, 1976.

A small infant in its mother's arms was reported to have been killed by hail at Fort Collins, Colorado, on July 30, 1979.

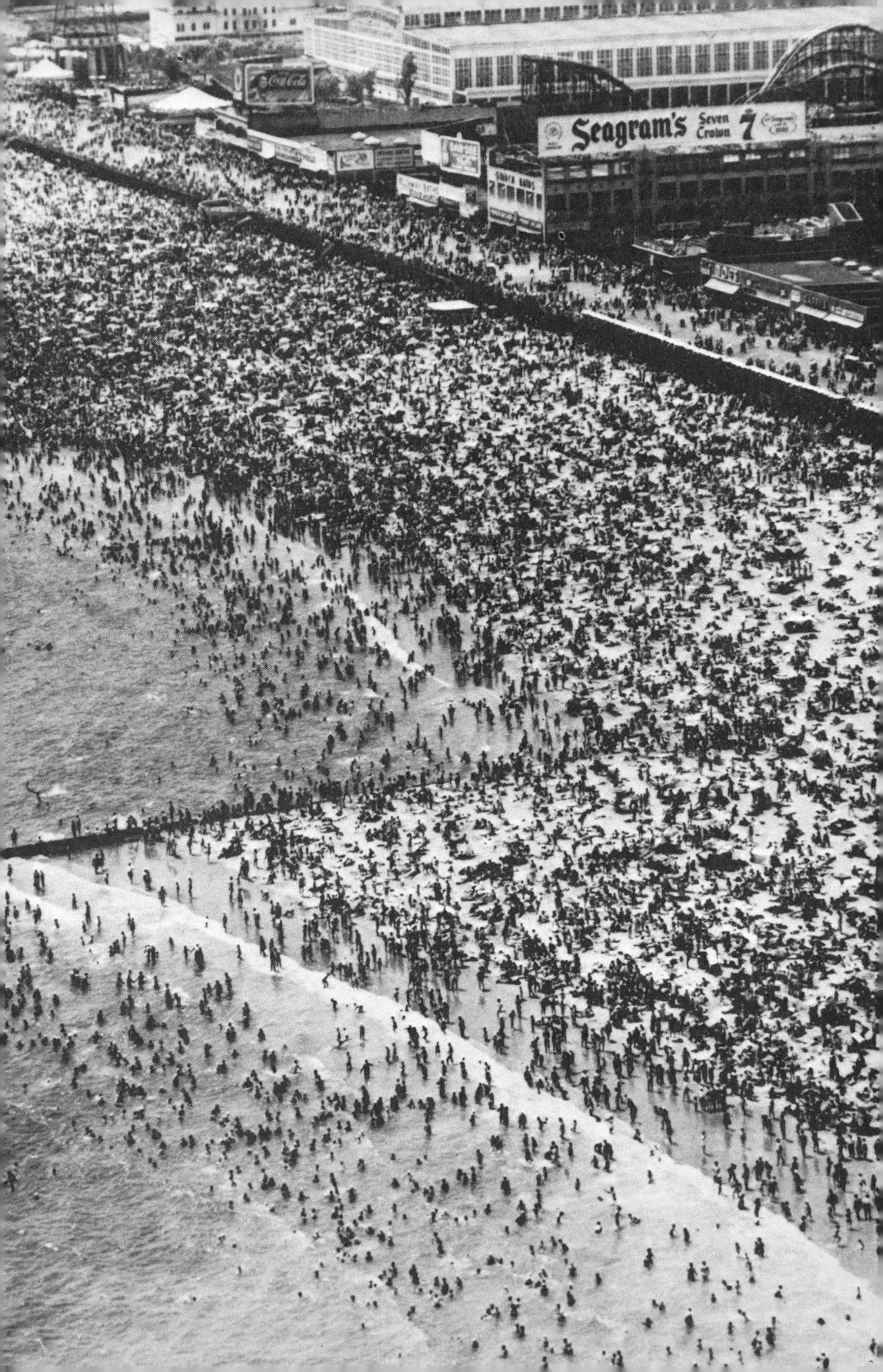
Coca-Cola
Seagram's
Seven
Crown
7

July

In this month is St. Swithin's Day,
On which if that it rain they say,
Full forty days after it will
Or more or less some rain distil.
— Poor Robin's Almanack (1697)

Opposite: Hot Sunday in summer draws over a million bathers to Coney Island, facing Atlantic Ocean in New York City. Wide World Photos.

When Sirius rises with the sun, mark the dog days well begun." From early July to early August, the brightest star of the sky rises at approximately the same time as the sun and accompanies our orb across the sky. The Romans reasoned that the combination of the radiation of Canincula, their dog star, and the sun produced the extreme heat and humidity characteristic of the July *dies canicula,* or dog days. We now know that the heat output of Sirius is about one ten-millionth part that of the sun, but the continued use of the phrase does give some an excuse for not expending too much human energy during the forty days from July 3 to August 11, when the temperature-humidity index is usually at its peak.

Another meteorological tradition of July concerns its often rainy nature in the more humid sections of the country. On the fifteenth is celebrated the feast of St. Swithin, the worthy bishop of Winchester in England, whose reburial in a more honored position a century after his death was delayed for forty days in 862 A.D. by a series of rainy days. Many countries in Europe have similar traditions, though each one names a different saint for the responsibility of initiating a long period of "falling weather," as rainy periods used to be called.

From June to July there is a marked northward shift of most features controlling the general circulation of the atmosphere in the northern hemisphere. The Pacific High continues its migration and occupies a mean axis across the eastern portion of the ocean along the latitude of Monterey, California. With its strength increased, it dominates the weather traffic over the northeast corner of the Pacific Ocean and the adjacent land, bringing a dry regime to southeast Alaska and the coastal strip south to California. The Aleutian Low joins the Siberian Low over northeast Asia and has a much reduced influence over weather conditions on the North American continent. The thermal low over the interior of Southern California, Arizona, and Nevada attains full summer strength; temperatures reach their maximum readings of the year and relative humidities fall to their lowest percentages over the Southwest.

In the Atlantic Ocean, the Azores–Bermuda High continues strong and occupies a northerly position with its west–east axis along the latitude of Cape Hatteras. It extends inland over the

Carolinas and exerts an important influence on temperature and humidity conditions over the eastern third of the United States. Some of our most uncomfortable summer weather occurs when the center of the Azores–Bermuda High lies in the western Atlantic Ocean close to the North American mainland. The northerly position of the high-pressure zone permits an easterly flow along its southerly periphery that carries low-pressure troughs from West Africa across the ocean to the West Indies. Some of these develop into tropical storms and hurricanes. The Iceland-Greenland Low occupies a position in southern Davis Strait between Greenland and Baffin Land and is much reduced as an influence on the weather traffic of the North Atlantic.

Storm tracks move north in July. In fact, there is no major path of storm centers within the United States. The only area of cyclone formation lies in the northern Plains, and these low-pressure centers head northeast, passing over Lake Superior and into southern Canada. The main continental storm track crosses central Canada along latitude 60°N and carries Alberta lows eastward over Hudson Bay to the Iceland–Greenland Low in Davis Strait; trailing fronts from these affect only the northern tier of states. Increasing tropical storm activity in the Atlantic easterly current directs disturbances from the waters near the Lesser Antilles west across the Caribbean Sea to Central America or northwest through the Yucatan Channel into the western Gulf of Mexico.

Summer heat reaches its greatest intensity in July. The 80°F (27°C) mean isotherm covers part of North Carolina, all of the Southeast except the higher mountain elevations, the Mississippi River Valley up to St. Louis, most of Arkansas, a good part of Kansas, and all of Oklahoma and Texas except the extreme west. All of the North has means of 70°F (21°C) or above except the border fringe from Montana eastward and the higher elevations of the Appalachians from West Virginia to Maine. In the Southwest, the 90°F (32°C) isothern makes an appearance over the contiguous portions of Arizona, Nevada, and California.

Temperature extremes for July: maximum 134°F (57°C) at Greenland Ranch, Death Valley, California, on July 10, 1913; minimum 10°F (−12°C) at Painter, Wyoming, on July 21, 1911.

July is much drier than June in the great grain-growing sections of the American heartland. Only isolated spots average above 4.0 in (102 mm). The area having more than this includes parts of interior New England and New York, then runs from a point near Buffalo southwest through the lower Ohio Valley to Arkansas, then south to extreme northeast Texas. All the territory from this northeast–southwest diagonal west to the crest of the Rocky Mountains lies in the 2–4-in (51–102-mm)-rainfall zone. Excessive rains fall along the Gulf Coast with the entire shoreline from Key West to Louisiana having 8.0 in (203 mm) or more.

July is a month of great activity on farms. The harvesting and threshing of small-grain crops progresses and is usually finished if favored by fair weather. Ample moisture is required for the development of corn; in many areas it is the most critical month for this crop. Winter wheat harvests usually become general by the first of the month as far north as the Ohio Valley, northern Missouri, and northern Kansas, and by July 20 in more northern sections. Harvest normally begins in the southern portions of the spring wheat belt about July 10. Cultivation of cotton is mostly completed during this month, with plants blooming and fruiting to the northern limits of the belt and picking begun in southern portions. Corn is coming into tassel or into silk in northern producing sections. Early potatoes are dug as far north as New Jersey and the Ohio Valley.

July

1 1792 *A true and particular narrative of the late tremendous tornado or hurricane at Phila. & N.Y. on Sabbath-Day* described a severe line squall and admonished Sabbath-breakers, since many young people were drowned while out boating on a Sunday.

2 1843 Alligator reported to have fallen at Charleston, S.C., on Anson Street during a thunderstorm.

3 1863 Battle of Gettysburg; great three-day encounter reached a climax with Pickett's Charge about 2:00 P.M.; weather partly cloudy, 87°F (31°C), wind SSW light, according to Prof. Jacob's records at Gettysburg College.

4 1776 Thomas Jefferson paid for his first thermometer and signed Declaration of Independence; weather was cloudy at 2:00 P.M., temperature 76°F (24°C), according to his "Weather Memorandum Book."

5 1916 Mobile's damaging early-season hurricane; wind 82 mi/h (132 km/h); barometer 28.92 in (97.9 kPa), tide 11.6 ft (3.5 m) above normal.

6 1928 Famous Potter, Neb., hailstorm; one stone measured 5.5 in (140 mm) diameter, 17 in (432 mm) circumference; weighed 1.5 lb (0.68 kg).

1936 Temperature of 121°F (49°C), highest ever recorded in North Dakota.

7 1911 Intermission in one of New England's severest heat waves: 90°F (32°C) or more from July 2 to 12 except 7–8; reached 106°F (41°C) in New Hampshire and Massachusetts, 105°F in Vermont.

8 1816 Frost in low places throughout New England on July 8–9.

9 1860 "Hot Blast" of torrid summer in Kansas; 115°F (46°C) at Lawrence and Ft. Scott.

1936 New York's all-time maximum of 106°F (41°C) at Central Park Observatory.

10 1913 Greenland Ranch in Death Valley, Calif., reported 134°F (54°C), long regarded as all-time U.S. maximum.

1936 Hottest day ever in East: 111°F (44°C) in Pa., 110°F (43°C) in N.J., 109° (43°C) in Md., 112°F (44°C) in W. Va.

11 1888 Mt. Washington, N.H.: heavy snow reached almost to base of mountain; Green Mountain peaks whitened.

12 1951 Worst flood ever inundated Kansas and Missouri; 16 in (406 mm) rain fell in three days on sodden ground; 41 drowned; $870 million loss; industrial Kansas City hard hit.

13. 1895 Cherry Hill Tornado in north New Jersey did $50,000 damage and killed three persons; funnel then descended at New York City in Harlem and Woodhaven areas where one more was killed; ended as waterspout in Jamaica Bay.

1977 Lightning strike on power line near Indian Point, N.Y., triggered massive 24-hour power blackout in New York City; looting resulted in billion dollar loss.

14 1936 Iowa's hottest afternoon, average maximum at 113 stations 108.7°F (42.6°C).

15 1980 Series of small tornadoes, hailstorms, and severe thunderstorms hit suburbs of Minneapolis; straight downrush winds did about $43 million damage, mostly in Dakota County; 100,000 customers without power .

15 1643 Gov. John Winthrop described small tornado at Newbury, Mass.: "Through God's mercy it did no hurt, but only killed one Indian."

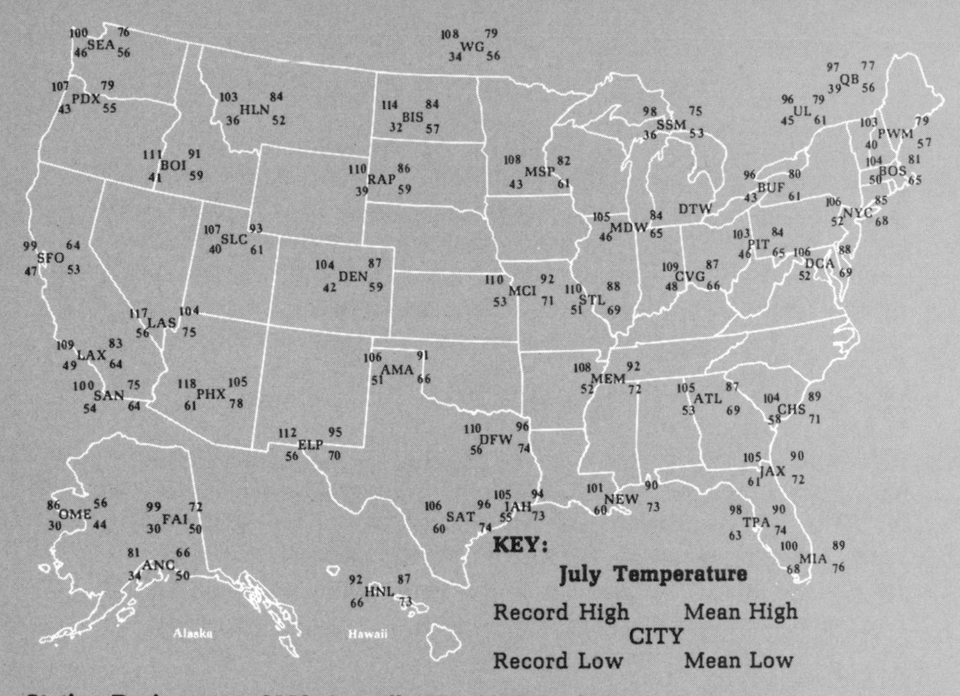

Station Designators: **AMA** Amarillo TX; **ANC** Anchorage AK; **ATL** Atlanta GA; **BIS** Bismarck ND; **BOI** Boise ID; **BOS** Boston MA; **BUF** Buffalo NY; **CHS** Charleston SC; **CVG** Cincinnati OH; **DCA** Washington DC; **DEN** Denver CO; **DFW** Dallas-Fort Worth TX; **DTW** Detroit MI; **ELP** El Paso TX; **FAI** Fairbanks AK; **HLN** Helena MT; **HNL** Honolulu HI; **IAH** Houston TX; **JAX** Jacksonville FL; **LAS** Las Vegas NV; **LAX** Los Angeles CA; **MCI** Kansas City MO; **MDW** Chicago IL; **MEM** Memphis TN; **MIA** Miami FL; **MSP** Minneapolis-St. Paul MN; **NEW** New Orleans LA; **NYC** New York NY; **OME** Nome AK; **PDX** Portland OR; **PHX** Phoenix AZ; **PIT** Pittsburgh PA; **PWN** Portland ME; **QB** Quebec QUE; **RAP** Rapid City SD; **SAN** San Diego CA; **SAT** San Antonio TX; **SEA** Seattle WA; **SFO** San Francisco CA; **SLC** Salt Lake City UT; **SSM** Sault Ste. Marie MI; **STL** St. Louis MO; **TPA** Tampa FL; **UL** Montreal QUE; **WG** Winnipeg MAN; **YC** Calgary ALB.

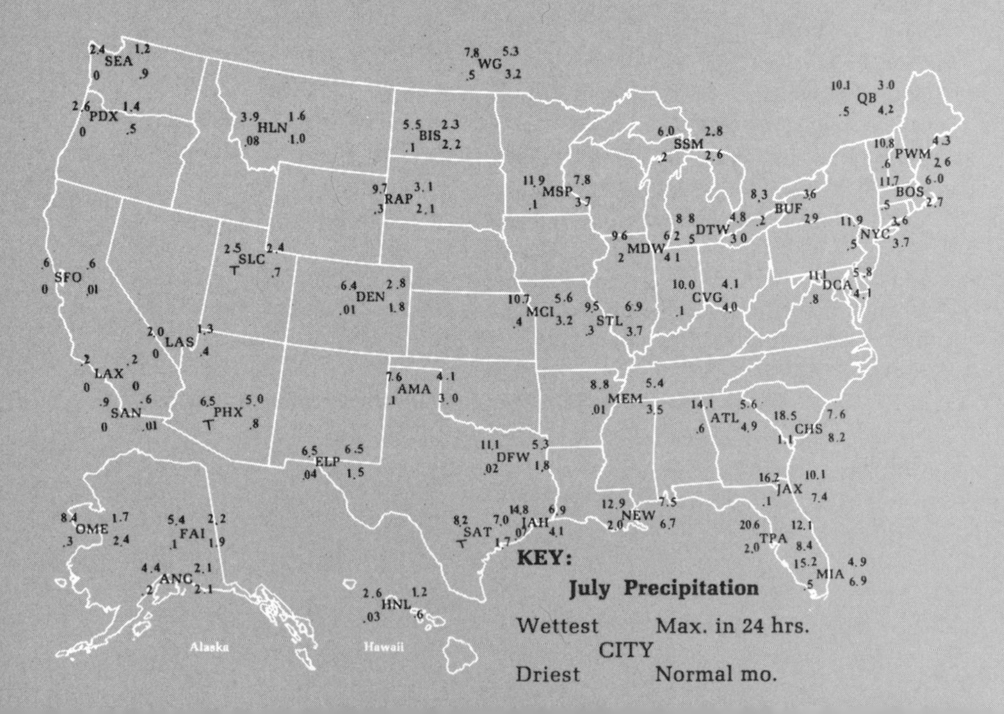

16 1945 First atomic bomb imploded at Alamogordo, N.M., under clearing skies after nighttime thunderstorm delayed the blast.

17 1941 Prolonged heat wave in state of Washington ended with numerous untimely thunderstorms that set 598 serious forest fires.

18 1942 Record deluge reported at Smethport in northern Pennsylvania; 30.7 in (780 mm) in six hours; 34.3 in (871 mm) in 24 hours.

19 1886 Third hurricane within month crossed northern Florida peninsula, Cedar Keys to Jacksonville, causing great damage.

20 1890 "Snow fell to an appreciable depth" during hailstorm at Calais, Maine.

1930 Washington, D.C., hit 106°F (41°C), the highest ever in the nation's capital.

1977 Johnstown, Pa., again flooded by 8.5-in (216-mm) cloudburst in 7 hours; 12 in (305 mm) fell nearby; 76 drowned; $424 million damage.

21 1975 New Jersey deluge of 10 in (254 mm) after rainy week flooded central state and Trenton; cut New York–Washington rail service for two days.

22 1858 Cambridge, Mass., thunderstorm: "generally considered as one of the most terrific within the memory of the present generation"; twenty lightning strikes in vicinity.

23 1788 George Washington's Hurricane; center passed directly over Mount Vernon as described by the General in his weather diary.

24 1886 Very dry season in Kansas ruined crops; no rain at Lawrence since June 26.

25 1891 Peak of torrid heat wave at Los Angeles with 109°F (43°C).

26 1819 Twin cloudbursts of 15 in (381 mm) struck almost simultaneously at Catskill, N.Y., and Westfield, Mass.; flash floods caused enormous erosion.

27 1926 Hurricane came inland near Daytona Beach, Fla.; $2.5 million damage in East Florida, including Jacksonville area.

28 1819 Bay St. Louis Hurricane in southern Mississippi, where Camille hit in 1969; small but intense storm considered "worst in 50 years"; U.S. cutter lost with 39 hands.

29 1905 Bridgeport Flood in southwest Connecticut; 11.32-in (288-mm) rain caused dam breaks; $250,000 damage in city.

30 1965 Portland, Ore., hit 107°F (42°C) to equal all-time maximum.

31 1976 Thompson Canyon Flood in northeast Colorado; stationary thunderstorm dropped 10 in (254 mm) in mountains; flash flood caught campers, 139 drowned with 5 missing; extensive structural and highway damage, estimated at $35.5 million.

Heat Waves

A period of abnormally hot weather continuing over an extended number of days has a great physiological and psychological, as well as economic, impact on the population. It requires a special designation in the meteorological vocabulary.

A heat wave was once defined as a spell of three or more consecutive days when the shade temperature reaches 90°F (32°C) on each day. But this lacked a consideration of the humidity factor so vital to human comfort. Thus, the temperature-humidity index (THI) was devised in the mid-1950s to give a numerical value to the relative comfort of a hot day. The criteria for the temperature and humidity relationship, of course, vary in different parts of the country. Generally speaking, a THI of 75 or more makes people uncomfortable and 80 or more makes everyone miserable.

A heat wave of long continuance tends to increase the incidence of respiratory and circulatory ailments with a resultant increase in mortality rates, as was demonstrated during the hot spell in the St. Louis area during July 1966.

Practically all large urban communities in the United States have experienced temperatures of 100°F (38°C) or higher unless they are favored by geography with a local cooling influence.

Along the coast of the Pacific Northwest from Cape Mendocino north, where upwelling cold water offshore tempers the prevailing west wind and creates foggy conditions, the following are the all-time maximums: Eureka, California, 85°F (29°C) and Tatoosh Island, Washington, 88°F (31°C). Astoria, Oregon, 8.5 miles (14 km) from the mouth of the Columbia River, once reached 100°F (38°C) when a hot airstream from the interior blasted down the valley. Inland cities have all reached the century Fahrenheit mark: Portland 107°F (42°C) and Seattle 100°F (38°C), but elevations in the mountains above 4000 ft (1219 m) generally have not attained three figures.

The higher mountain exposures of the Intermountain region and the Rocky Mountains are also favored with low summertime maximums: Flagstaff, on the elevated tableland of Arizona at 6993 ft (2132 m), has a peak reading of 96°F (36°C), and Alamosa, at 7536 ft (2297 m) high in the San Juan Valley of southern Colorado, has experienced a maximum of only 91°F (33°C). No 100°F temperatures have been registered in Yellowstone Park: Lamar Ranger Station, high in the northeast section, has reached 97°F (36°C), and Mammoth Springs, 6241 ft (1902 m), once sweltered at 95°F (35°C).

From the Rockies eastward to the western slopes of the Appalachians lies the home base of American heat waves. All locations except those on lee exposures of the Great Lakes have reached 100° F (38°C). The cooling effect of the lake waters is well illustrated by some cross-lake contrasts: Muskegon, Michigan, on the east side of Lake Michigan, has attained only 99°F (37°C), while Milwaukee, Wisconsin, directly across on the western side, has soared to 105°F (41°C). In the eastern Lakes the figure of 99°F has been the highest ever reached at Buffalo, New York, and Erie, Pennsylvania, in over 100 years of official thermometer watching.

Some locations in the highlands of the Appalachians, too, have escaped the century mark. Asheville, in western North Carolina at 2203 ft (672 m), once registered a high of 99°F (37°C), as did Elkins in West Virginia, at 1940 ft (591 m). In the north, the official thermometer atop Mt. Washington, under surveillance since 1932, shows a peak reading of 72°F (22°C) at the 6267-ft (1910-m) weather station elevation.

All locations along the Gulf Coast from south Texas around to northern Florida have experienced a searing, humid 100°F (38°C) or more as a result of land winds blowing from interior heat waves, but the cities on the fringe of the Florida peninsula are normally cooled most of the time by ocean breezes. Their maximums are: Tampa, 98°F (37°C); Key West, 97°F; and Miami Beach, 98°F.

Continental heat waves on occasion can engulf the entire Atlantic seaboard except northeast Maine and drive the mercury over the century mark. Only exposed locations with overwater trajectories, such as Cape Hatteras, North Carolina, with a maximum of 97°F (36°C), and Block Island, Rhode Island, with 95°F (35°C), have not touched the century mark; but Nantucket Island, 30 mi (48 km) at sea, recently saw its all-time maximum raised to 100°F on Hot Saturday, August 2, 1975.

North of Portland, Maine, the cool waters of the Gulf of Maine usually keep the natives and summerfolk in the comfort zone, but the record books were changed here, too, on Hot Saturday when Portland reached 103°F (39°C) and Brunswick 104°F (40°C). The northeasternmost corner of the country, at Eastport, shows an absolute maximum of 93°F (34°C) in a record stretching back to 1873, and Caribou in the extreme north of the Down East State has reached a maximum of 96°F (36°C) several times.

Historic Heat Waves

Great Plains, 1901 — July 1901 earned the reputation of being the hottest month ever experienced in the central part of the country since the collection of state weather records was systematized in the 1880s. At Lawrence, Kansas, it was the warmest month since 1867, when record-keeping at the University of Kansas was begun, and it had been preceded by a record hot June. The mercury in July reached 90°F (32°C) on every day of the month and touched 100°F (38°C) or more on 21 days (13 of these being consecutive, from July 8 to 25), and soared to 104°F (40°C) on nine days in a row. The peak of 108°F (42°C) on July 24 set a new record maximum for the 34 years of observations at Lawrence. The state maximum in July 1901 in Kansas was 112°F (44°C) at Phillipsburg.

In Iowa the average temperature for July 1901 for the state as a whole was 82°F (28°C), or 6°F (3.3°C) higher than any previous July figure. The absolute state maximum of 113°F (45°C) represented an increase of 3°F (1.7°C) over any reading previously registered. The town of Villisca had 19 consecutive days with 100°F or more, and at West Bend it was said to have been "the hottest for 50 years," or since the settlement of the area. The state weather director at Des Moines, in commenting on the long-sustained heat wave with century readings, thought that "they will stand for comparison in future years, and it may be hoped that the new record will remain unbroken during the coming century."

Great Plains, 1930s — Though hot spells came and went during the next three decades, there were no torrid spells or continuous heat waves such as were experienced in 1900 and 1901 until the 1930s. This was the famous Dust Bowl period, when drought and heat combined with the Great Depression to bring on the worst human catastrophe, aside from the Civil War, that the nation has ever experienced. During the period from 1930 to 1936, all Julys were above long-term normals in Iowa and Kansas, and of the seven Augusts, five in Kansas and six in Iowa were above normal. It was the combination of long-continued drought and an increasingly intense degree of summer heat that spelled economic doom for the area as the 1930s progressed.

The outstanding heat terms came in the summers of 1930, 1934, and 1936. Generally speaking, records for absolute maximums set in 1930 were exceeded in 1934 and again topped in 1936.

C. D. Reed, the head of the Iowa Weather and Crop Bureau, made some extensive studies of past heat conditions in his state. Concerning July 1936, he concluded:

> Comparing the mean temperatures at individual stations in all Julys back to the beginning of records in 1819, with the July 1930 mean temperatures in these localities, there is ample margin in favor of 1936 to take care of all possible differences due to location and exposure of instruments and methods of observation, and still leave July 1936 well in the lead as the hottest July in 117 years.

The hottest afternoon ever experienced in Iowa came on July 14, 1936, when the average maximum temperature of 113 stations throughout the state was 108.7°F (42.6°C). Kansas City is the largest city that borders on the Plains, and it had a torrid time in the summer of 1936: The mercury soared to 100°F or higher on 53 days. The all-time maximum in Kansas City history came with 113°F (45°C) on August 14.

The summer of 1936 set marks for absolute maximums that still stand unsurpassed in 16 states, mostly located on the Great Plains, though some were in the Ohio Valley and Middle Atlantic States. The extreme figures reached were: 121°F (49°C) in Kansas and North Dakota and 120°F (49°C) in South Dakota, Texas, Arkansas, and Oklahoma. These are the only instances of readings of 120°F or more outside the Southwestern desert triangle of California, Nevada, and Arizona. (There is one doubtful reading of 120°F in Oklahoma in 1943.) In July 1936 a large number of stations on the Great Plains attained 115°F (46°C) or more: in North Dakota, 22 stations; South Dakota, 20; Nebraska, 24, and Kansas, 41.

Northeast, 1955 — Never did so many Americans swelter under such high temperatures through such a long summer as did the residents of the large industrial cities of the Northeast in July and August 1955. Over much of the territory from the central Great Plains eastward to the Atlantic Coast, but excluding the Gulf States, persistent anticyclonic conditions, almost without a break, maintained temperatures 4°F (2°C) or more above normal. The heat wave was noted for: (1) its persistent nature, (2) extreme high means, (3) lack of new absolute maximums, and (4) absence of drought.

The industrial cities of the Midwest and Northeast, with their great concentrations of population, suffered most under the persistent high temperature–humidity index conditions. July 1955 brought Chicago, Detroit, Cleveland, Hartford, New York, Philadelphia, Baltimore, and Washington their warmest month of record. In addition to those listed above, the following had their greatest number of torrid days with temperatures of 90°F (32°C) or above in a single month: Minneapolis, Lansing, Indianapolis, Albany, and Boston.

The vast metropolitan area of New York City endured its warmest calendar month, too, in July 1955. The mean of 80.9°F (27.2°C) topped that of the previous warmest month, July 1952, by a small fraction. These are the only two months in National Weather Service records for New York's Central Park station to exceed 80°F (27°C).

August continued the summer heat wave of 1955 over much of the same Eastern area, with departures from normal of 4F (2.2C) degrees and more being common. Cleveland followed its record July with an August hot spell that broke all marks for that month. Chicago reached 90°F or more on a record number of days in the summer of 1955.

Southwest in the 1950s — The desert portion of the country, where insolation is at a maximum, experienced its most extended period of extreme heat during the last half of the decade of the 1950s. A preview of the torrid events to come was staged in late summer 1955, when an eight-day scorcher in Southern California put together the greatest number of days of 100°F or more in the long history of the downtown Los Angeles weather office. During this period, from August 31 to September 7, the mercury pushed up to a peak of 110°F (43°C) on the first of September, the hottest ever to visit the City of the Angels since official records were begun in 1877.

The long term of unrelenting heat commenced anew in April 1956 and held the area in its grip until November 1960. Every month during this extended period averaged above normal at the Los Angeles International Airport weather station, which lies only 3 mi (4.8 km) from the Pacific Ocean. The climax of the heated period came in the summer of 1959, when June, July, August, and September each exceeded former records. Averaging 7.5F (4.2C) degrees above normal, July was the warmest calendar month in the 16-year span of records at the airport.

Inland locations, too, experienced heat in its most persistent form. Yuma, on the Arizona-California border, established new high averages in both June and July 1959 — the latter, with 96.7°F (35.9°C), was the warmest month ever for the normally warmest urban location in the United States. The mercury soared past the century mark every day of the month; the average daily maximum was 109.4°F (43°C); and the peak reading of 118°F (47.8°C) came on July 10.

Down in Death Valley, the super-heated air reached a figure of 125°F (52°C) at Cow Creek for California's highest that summer; North Las Vegas in Nevada rose to 120°F (49°C). The bulge of hot air also spread northward over the entire Intermountain region. For instance, Pocatello, Idaho, had 18 consecutive days with 90°F (32°C) or more — a new record for a heated term there.

The next summer brought a continuation of the climate trend and made some additional marks of its own. At Yuma, July 1960 averaged the second hottest such month, only 0.9F (0.5C) degrees below the previous July.

Yuma: The Hottest Urban Area in the United States

Highest temperature:	123°F (51°C), Sept. 1, 1950
Highest daily mean:	103.5°F (40°C), July 31, 1957
Highest daily minimum:	89°F (32°C), July 24, 1959
Highest monthly mean:	96.7°F (36°C), July 1959
Highest annual mean:	76.3°F (25°C), 1958

Consecutive days:

Maximum 90°F (32°C) or more	153 in 1973
Maximum 100°F (38°C) or more	101 in 1937
Maximum 110°F (43°C) or more	14 in 1955
Minimum 80°F (27°C) or more	30 in 1959

Number of days in month:

Minimum 85°F (29°C) or more	16 in July 1959

Texas Heat in 1980 — All records for duration of high temperatures were broken over much of Texas during the first half of summer 1980. Commencing on June 23 and continuing through August 3, a period of 42 days, the maximum thermometer reading at the Dallas–Fort Worth Airport rose to 100°F (38°C) or more. The extreme readings came on June 26 and 27, when the temperature peaked at 113°F (45°C) to exceed the old record by one degree. The month of July, with a mean of 92°F (33°C), was the warmest month ever for the north-central Texas urban complex, and the three months of summer 1980 were the warmest with a mean of 89.2°F (31.8°C).

It was even more torrid in the Wichita Falls area, along the Red River, where the mercury exceeded 110°F (43°C) every day from June 23 to July 3. The highest came on June 28 with a reading of 117°F (47°C), an all-time record for Wichita Falls. The extreme reading in all of Texas was 119°F (48°C) at Weathersford, west of Fort Worth; this was only one degree short of the Texas record set in the torrid summer of 1936.

August

Hurricane Alert Calendar

May	**not today**
June	**radios in tune**
July	**all stand by**
August	**beware you must**
September	**time to unlimber**
October	**not yet over**
November	**remember**

— Variation on an old Florida proverb

Residents of Gulf Coast localities and the entire Atlantic seaboard should turn their eyes toward the tropics in August if they have not already done so. Although in July there is a possibility of a tropical storm's exploding into a full-fledged hurricane, it becomes a probability in August. July and August were months added to the original Roman calendar, being named by and after two Caesars, Julius and Augustus. Weatherwise folk think the climatic reputations of the Caesars might have been better regarded if other months of the year had been selected to honor them.

The tropical waters of the Atlantic Ocean, stretching all the way from West Africa to the West Indies, become a breeding ground for cyclonic disturbances that have the potential of growing into August hurricanes. The outstanding storm of our early history, the Great Colonial Hurricane, struck New England on August 25–26, 1635; and the greatest in recent times reached full maturity on August 17, 1969, when Camille smashed ashore in Mississippi and then loosed unprecedented deluges of rain as far north as Virginia. A tropical storm, or even a hurricane, may not be all bad at this season, since it brings beneficial rainfall at a season when the precipitation supply may be irregular. The long continuance of the great drought in the Northeastern States in the early and mid-1960s was attributed to the complete absence of tropical storms and attendant rainfall during that period. It has been estimated that tropical storms contribute about 15 percent of the annual rainfall in the Gulf States, and in some years much more.

The Pacific High is at its greatest expanse in August. Its mean axis moves north to about 38°N along the latitude of San Francisco, and its center has increased in strength to about 30.27 in (102.5 kPa). The northeast extension reaches inland over British Columbia and the panhandle of Alaska, providing an effective block to storm movement southeastward from Alaska proper. A mean low center exists in the Bering Sea and is attached to the Siberian Low, while the thermal low in the Southwest remains at full summer strength.

In the Atlantic Ocean, the Azores–Bermuda High occupies a position across the entire ocean, with the center of highest pressure along 35°N, the latitude of Cape Hatteras, and extends inland over the coastal plain from Maryland to Florida. In August, it continues its July role of pumping heat and humidity into the northern states. Over the subpolar waters, a low-pressure zone extends from Baffin Land east to Icelandic shores and is at its lowest strength of the year as an influence on the movement of weather systems.

The mean entrance point of the Pacific jet stream to the continent moves north to the latitude of Tacoma, Washington, where it heads east-northeast across the extreme southern Prairie Provinces of Canada, passing north of the Great Lakes, and leaving the continent over Newfoundland. Its strength and influence on United States weather reach a low ebb in August.

The main continental storm track migrates poleward, too, in August; the most frequented path for Alberta disturbances lies close to 60°N over central Hudson Bay. General storms are infrequent within the conterminous 48 states except along the Atlantic Coast. The tropical easterlies move north, with a marked increase in storm activity there. Some major disturbances originate near West Africa, in the vicinity of the Cape Verde Islands, as low-pressure troughs or waves in the easterly current are carried west across the warm waters of the southern North Atlantic Ocean. Upon approaching the Greater Antilles, two paths may be followed: One continues west over the Lesser Antilles into the Caribbean Sea and on to Central America and Mexico; the other veers north of Puerto Rico and Cuba, either passing through the Florida Straits into the Gulf of Mexico or recurving to the north along the Atlantic seaboard.

In the United States, the primary anticyclone track across the northern border has increased traffic. The northern Plains continue as one of the most anticyclonic regions of the entire Northern Hemisphere. The predominant track of polar and arctic highs in North America undergoes an interesting westward shift from July to August. They now cross the Beaufort Sea, Mackenzie Valley, and Prairie Provinces as they will during the coming winter. Their frequency, however, is low. Warming of Hudson Bay puts an end to the development of anticyclones in this region, although some polar highs still form in nearby Manitoba.

The rainfall-distribution pattern of August generally resembles that of July over the eastern two thirds of the country. A difference lies in the lower Mississippi Valley, where the 4.0-in (102-mm) line has retreated to central Louisiana from its July position near St. Louis. An isolated area of more than 4.0 in appears in parts of Missouri, Iowa, and the middle Missouri Valley, between Kansas City and Omaha, and in adjacent parts of northern Wisconsin and Minnesota. The major development is the expansion of the moderate rainfall area in southern Arizona, over much of New Mexico and central Arizona, with patches in excess of 4.0 in. All of the Pacific States and Nevada remain in the area of less than one inch (25 mm), with the exception of the Cascades and the coastal mountains of Oregon and Washington.

The temperature-distribution pattern, too, is similar to that of July. The principal difference appears in the North, where it is a little cooler. The 70°F (21°C) mean isotherm, found only in the lowlands of New England and New York, has retreated south to central Michigan, Wisconsin, and Minnesota, and then runs along the border between the two Dakotas. The 90°F (32°C) or more area in the Southwest remains about the same size. The 60°F (16°C) line has advanced to the California coast south of San Francisco, although the coastal and mountain area to the north still continues below 60°F.

Temperature extremes for August: maximum 127°F (53°C) at Greenland Ranch, Death Valley, California, on August 12, 1933; minimum 5°F (−15°C) Bowen, Montana, on August 25, 1910.

In the higher elevations of the more Western states, winter wheat harvest becomes general in August, while that of spring wheat starts in the north-central portions of the country. Corn usually requires ample moisture until at least the middle of the month. In the upper Mississippi Valley the crop usually reaches the roasting-ear stage about the middle of August, or a little later; in the Southern states corn has fully matured in the early part of the month. Picking begins in the southern Cotton Belt during the first 10 days of August and advances by the latter part of the month to southern North Carolina, northern Alabama, and southern Oklahoma.

August

1 1954 Snow cover of 16 in (41 cm) remained on Mt. Rainier in Washington at 5550 ft (1692 m) elevation after big snow season.

2 1975 "Hot Saturday" set all-time heat records in eastern New England: 107°FF (42°C) in Massachusetts, 105°F (41°C) in Maine and New Hampshire, 104°F (40°C) in Rhode Island.

3 1970 Hurricane Celia struck the Corpus Christi area; "the most damaging ever on the Texas coast," with 11 deaths and $454 million loss.

4 1961 Spokane's all-time maximum of 108°F (42°C).

1980 Dallas–Ft. Worth's string of 42 consecutive days with maximum of 100°F (38°C) ended; record hot summer, mean 89°F (31.7°C); new maximum record of 113°F (45°C).

5 1843 Spectacular cloudburst and tornado on Chester Creek near Philadelphia: 16 in (406 mm) fell in three hours; at Newark, N.J., deluge of 15 in (381 mm) measured by bucket survey when rain gauge spilled over.

6 1918 Philadelphia's most uncomfortable day with 100°F (38°C) and temperature-humidity index of 79.4.

7 1918 New York City's warmest day and night, with 102°F (39°C) maximum and 82°F (28°C) minimum; Philadelphia reached all-time maximum of 106°F (41°C); Washington 105.5°F (41°C) on August 6.

8 1882 August snowstorm on Lake Michigan when "thick cold cloud burst on decks, covered them with snow and slush 6 inches deep"; snow showers observed at shore points that day.

9 1878 New England's second most deadly tornado struck Wallingford, Conn.; 34 killed; 30 houses completely destroyed.

10 1856 Isle Derniere (Last Island) disaster off Louisiana coast; storm tide drowned 140 vacationers when 5 ft (1.5 m) wave swept over low island during hurricane.

11 1944 Burlington, in northern Vermont, hit 101°F (38°C) for all-time maximum record.

12 1778 Rhode Island Hurricane prevented impending British-French sea battle; caused extensive damage in southeast New England.

13 1831 Blue sun observed widely in South; thought to have presaged Nat Turner's slave uprising; phenomenon continued several days.

14 1936 Kansas City's highest reading ever: 113°F (45°C).

15 1787 Four-state tornado outbreak in New England over Connecticut, Rhode Island, Massachusetts, and New Hampshire; Wethersfield, Conn., hard hit.

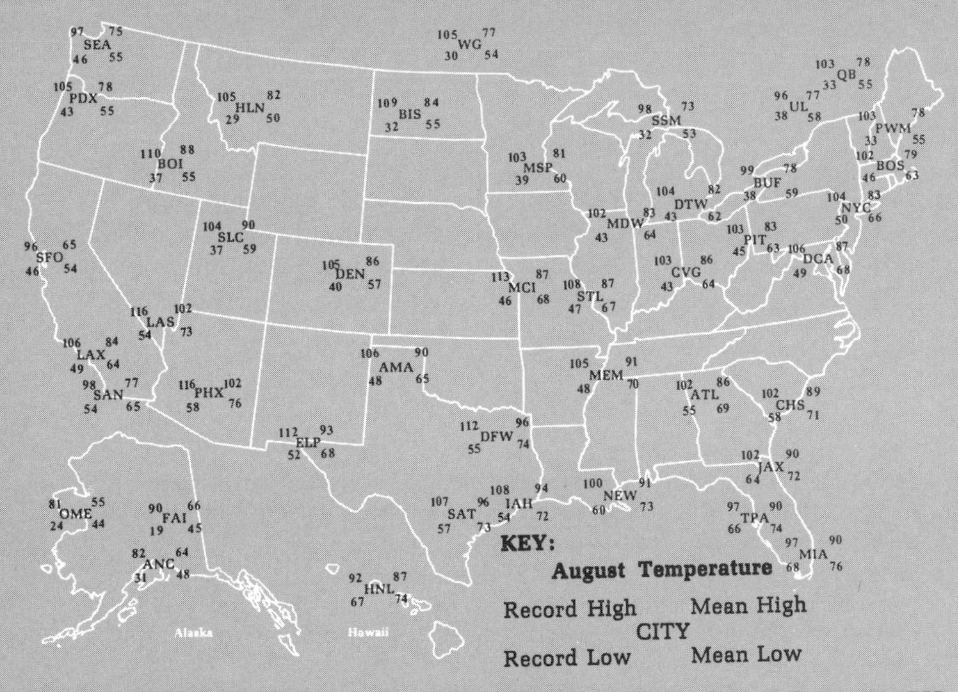

Station Designators: AMA Amarillo TX; **ANC** Anchorage AK; **ATL** Atlanta GA; **BIS** Bismarck ND; **BOI** Boise ID; **BOS** Boston MA; **BUF** Buffalo NY; **CHS** Charleston SC; **CVG** Cincinnati OH; **DCA** Washington DC; **DEN** Denver CO; **DFW** Dallas-Fort Worth TX; **DTW** Detroit MI; **ELP** El Paso TX; **FAI** Fairbanks AK; **HLN** Helena MT; **HNL** Honolulu HI; **IAH** Houston TX; **JAX** Jacksonville FL; **LAS** Las Vegas NV; **LAX** Los Angeles CA; **MCI** Kansas City MO; **MDW** Chicago IL; **MEM** Memphis TN; **MIA** Miami FL; **MSP** Minneapolis-St. Paul MN; **NEW** New Orleans LA; **NYC** New York NY; **OME** Nome AK; **PDX** Portland OR; **PHX** Phoenix AZ; **PIT** Pittsburgh PA; **PWN** Portland ME; **QB** Quebec QUE; **RAP** Rapid City SD; **SAN** San Diego CA; **SAT** San Antonio TX; **SEA** Seattle WA; **SFO** San Francisco CA; **SLC** Salt Lake City UT; **SSM** Sault Ste. Marie MI; **STL** St. Louis MO; **TPA** Tampa FL; **UL** Montreal QUE; **WG** Winnipeg MAN; **YC** Calgary ALB.

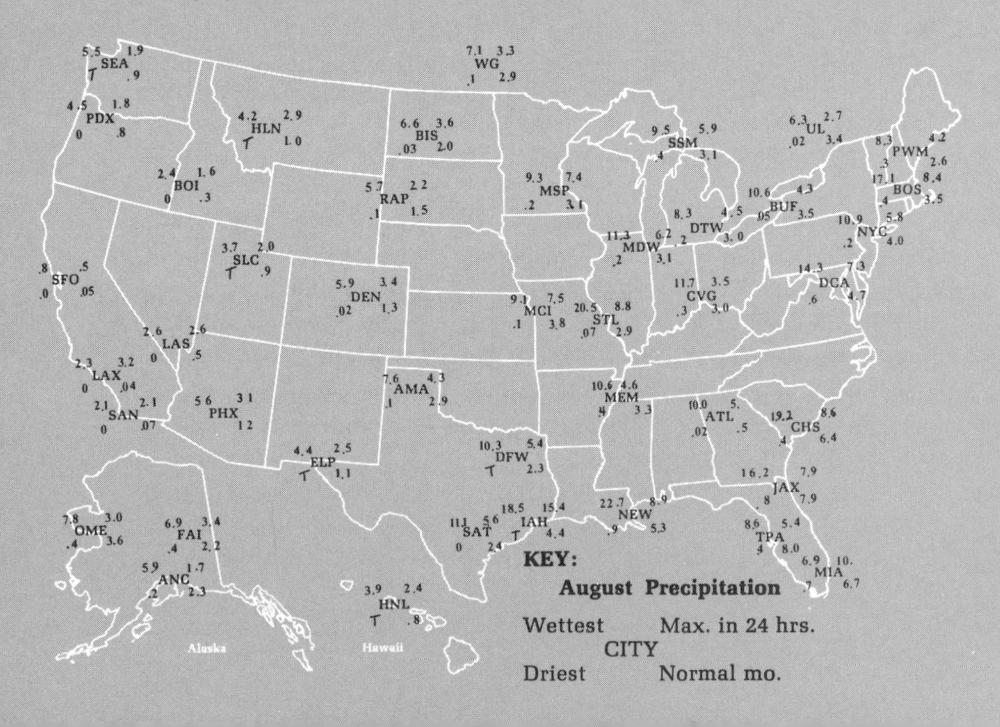

16 1777 Battle of Bennington; start delayed a day by rain, enabling Vermont militia to arrive for start, but rain delayed British reinforcements; Americans won victory by defeating the two enemy forces, one at a time.

17 1969 Hurricane Camille made landfall on Mississippi coast, "severest ever to strike populated area in U.S."; winds of 200 mi/h (322 km/h); tide 24 ft (7.3 m); caused 144 deaths and $1280 million damage; later floods in Virginia drowned 113 more persons.

18 1779 Hurricane eye, passing over New Orleans, was described by William Dunbar, the pioneer weather observer of Louisiana.

19 1788 Spectacular small hurricane caused great forest havoc along narrow track over interior from New Jersey to Maine; repetition today would cause extreme disaster in now built-up area.

20 1910 Big blowup of Idaho fires during driest month in history; 1736 fires, three million acres burned; 78 fire fighters and 7 others died; town of Wallace half consumed.

21 1883 Tornado at Rochester, Minn., 31 killed.
1918 Tornado at Tyler, Minn., 36 killed.

22 1816 Damaging frosts in low places from interior New England to North Carolina; corn crop shortened.

23 1933 Chesapeake–Potomac Hurricane moved over Norfolk and Washington; tide 7 ft (2.1 m) above normal flooded business section of Norfolk; damage in Maryland estimated at $17 million.

24 1906 Extreme cloudburst at Guinea, Va.: 9.25 in (235 mm) in 40 minutes.

25 1635 Great Colonial Hurricane; same physical stature of 1815 and 1938; track parallel to 1944; tide rose 20 ft (6 m); vivid accounts written by Gov. Winthrop and Gov. Bradford.

26 1949 Severe Delray Hurricane in Florida; wind 153 mi/h (246 km/h); barometer 28.17 in (95.4 kPa).

27 1964 Hurricane Cleo battered Miami and south Florida area; first direct hit since 1950; 135-mi/h (217-km/h) gusts; barometer 28.57 in (96.7 kPa); $125 million damage.

28 1971 Tropical storm Doria caused high floods in central New Jersey and southeast Pennsylvania.

29 1965 Cool wave brought 2.5-in (6-cm) snow atop Mt. Washington for August record; 25°F (−4°C) in Vermont, 39°F (4°C) on Nantucket; earliest freeze of record at many stations.

30 1776 Washington took advantage of heavy fog to evacuate Long Island after defeat; British fleet with adverse winds could not intervene.

31 1954 Hurricane Carol swept eastern New England; Providence inundated; 60 lives lost; $450 million damage.

Drought

Nothing is less inspiring to man than a drought. Many Americans live in areas of marginal water supply, sometimes having adequate rainfall, sometimes experiencing a serious deficiency. Once the Great Plains were known as "The Great American Desert," but man's ingenuity has partially subdued this vast region, and in most seasons its lands yield a bounteous harvest of crops, though the threat of drought is always present.

A period of abnormal moisture deficiency of sufficient length may cause a serious hydrologic imbalance resulting in crop damage, water supply shortage, and economic disruption. Dry spells occur sporadically in all parts of the country, but the term *drought* is reserved for periods of moisture deficiency that are relatively extensive in both time and space. A moisture shortage that may be termed a drought in one region may not be considered so in another. In years past, a period of 21 days with but 30 percent of normal rainfall was considered a drought.

Drought Index — A more sophisticated index of meteorological drought was developed by Wayne Palmer of the National Weather Service. The general concept is one of supply and demand: Supply is represented by precipitation and stored soil moisture; demand is the combination of potential evaporation and transpiration (evapotranspiration) and the amount of water needed to recharge the soil moisture and the runoff required to keep rivers, lakes, and reservoirs at a normal level. The results of this water-balance accounting produce either a positive or a negative figure that is then weighed by a climatic factor. The final product is an index that expresses the abnormality for that particular place in regard to moisture for a long-term period. If the index lies between +2 and −2, conditions are considered normal. If between −2 and −3, moderate drought exists; if between −3 and −4, severe drought is indicated; and below −4, extreme drought prevails.

In addition, a second index provides a short-term measure of the degree to which moisture requirements of growing crops were met during the previous week only. The Crop Moisture Index is computed from average weekly values of temperature and precipi-

Dust drift piles against a farmyard in the Dust Bowl years on the southern Great Plains. U.S. Department of Agriculture (USDA) photo.

tation, which give the potential moisture demand. Taking into account the previous moisture condition and current rainfall, the index determines the actual moisture *loss*.

If potential moisture demand exceeds available moisture supplies, the Crop Moisture Index gives a negative value. But if current moisture meets or exceeds demand, the index is positive. Charts of the Drought Severity Index and the Crop Moisture Index are published weekly during the crop season in *The Weather and Crop Bulletin,* issued jointly by the Department of Agriculture and the National Weather Service in Washington.

Historic Droughts — The crops of the Pilgrims were threatened by lack of rain in their first summer, 1621, when none fell for 24 days; and again, in 1623, another dry spell endangered their corn. During the severe droughts of 1749, 1761, and 1762 in eastern New England, fields of grain caught fire and dwellings were consumed. Very dry years occurred in 1805 and 1822 in the Eastern States. The summer of 1854 brought widespread drought and crop failures from New York to Missouri, and the year 1860 was notorious in newly settled Kansas for drought, heat blasts, and dust storms.

In California the record dry season of 1850–51 produced only 5.0 in (127 mm) of rain at Sacramento and only 7.4 in (188 mm) at San Francisco, or about 33 percent of normal. The two extremely dry seasons of 1862–63 and 1863–64 in Southern California caused the death of thousands of cattle and put an end to the grazing industry of the region.

After some 20 years of varying, but mostly adequate, rainfall on the Great Plains, drought struck in 1881 and continued endemic until 1887, when a combination of severe winters and droughty summers plus overproduction caused the collapse of the Cattle Kingdom. The following growing seasons were spotty, and economic distress was widespread. A massive exodus of people from Kansas and Nebraska commenced. A historian of the period has written: "Fully half of the people of western Kansas left the country between 1888 and 1892. Twenty well-built towns in that part of the state were reputed to have been left without a single inhabitant." A current saying was: "In God we trusted, in Kansas we busted." Things became worse rather than better when extreme drought prevailed in 1893, 1894, and early 1895, the Drought Severity Index reaching a low of −4.97.

Serious droughts threatened again in 1910, 1911, and 1913, but they were relatively short-lived.

The Great Plains Drought in the 1930s — The greatest disaster in American history attributable to meteorological factors occurred on the Great Plains in the decade of the 1930s and has been known popularly ever since as the Dust Bowl Days. Droughty conditions began in the latter half of 1930 and continued into the first half of 1931, when most of the northern and southern Plains experienced a severe moisture shortage. Thereafter, in every year until the end of the decade, some part of the region was affected by serious drought; and in 1934 and 1936 the entire region from Texas to Canada was scourged. Between August 1932 and October 1940, the drought severity index in western Kansas was continuously below normal, and in 38 of these months extreme drought prevailed. Low points were reached three separate times: −5.96 in August 1934, −5.55 in August 1936, and −5.55 in October 1939.

The name *Dust Bowl* applied only to the most seriously affected areas of the south-central Plains, which included the contiguous parts of Colorado, New Mexico, Texas, Oklahoma, and Kansas. At its greatest extent, in the winter of 1935–36, the Dust Bowl covered 50 million acres. For the previous three years, Dalhart, Texas, close to the center of the area, averaged 11.08 in (281 mm) of rainfall per year, or 58 percent of normal. The stricken area began to shrink in 1937, and by early 1938 was down to about 9.5 million acres, or about one fifth of its original size; but extreme drought again in 1939 expanded it somewhat. Good rains came finally in 1940 and early 1941 to put the area back into agricultural production, just in time to meet the extraordinary demands of World War II on the nation's productive capacity.

Drought in the 1950s — Deficient precipitation again made itself felt early in the 1950s, and dust storms once again swirled across the Great Plains. At Amarillo, in northwest Texas, "the worst drought of record" prevailed from mid-1952 to early 1957, when the annual precipitation averaged only 12.55 in (320 mm), or 61 percent of normal, for that lengthy period. The summer of 1954 was the hottest and driest combination in a large zone from the southern Plains eastward across the Deep South to the Atlantic Ocean. Crop yields were down 50 percent in much of the South and Southwest. Next year, vast dust storms returned in springtime, and the area of less than 50 percent crop yield migrated westward to center in Arkansas, Oklahoma, and Texas.

In western Kansas the Drought Severity Index dropped into negative figures in May 1952 and did not emerge on the plus side until March 1957, a period of 57 months compared to the 99-month duration in the 1930s. But the index dropped to an all-time low for

the region in September 1956, with a figure of −6.2. As early as summer 1953 there was severe drought, and 31 counties of western Kansas were declared disaster areas. Conditions became worse in the autumn of 1954, then spring rains produced some relief in 1955, but by the end of summer conditions again deteriorated. "The moisture was dismal all during 1956. The spring was dry and the summer and the fall were drier. The total runoff for the year was the lowest of record," reported *The Weather and Crop Bulletin.* The index value went into the extreme-drought category in May 1956 and remained there until copious rains came in March 1957.

Northeast Drought, 1961–66 — Abnormally dry conditions over a sizable area in the Northeast first appeared in September 1961, and by the end of October mild-to-moderate drought was established over a narrow belt from northern Virginia to southern Vermont. In 1962 the drought reached moderate-to-severe proportions during July and August, but adequate precipitation over the winter eased most water shortages by spring 1963.

Serious concern among public officials over the effects of the prolonged period of predominantly dry weather appeared again during October 1963, the warmest and driest October of record at many stations. Water shortages developed, and by the end of the month New York City's reservoirs dropped to about 30 percent of capacity. Forest fires were widespread during the autumn.

After six months of relatively wet weather and improved runoff, dry weather returned again in May 1964. By September many areas in the Northeast were in the extreme drought category and reservoirs stood at record low levels. During 1965 the core of the drought area remained in the Hudson Valley, but its influence widened and spread southward. The three summer months of 1965 gave New York City only 37 percent of normal rainfall and the entire year brought only 50 percent. The extreme drought area now reached from southern New England to central Virginia.

Moderate precipitation in January and February 1966 only

Opposite: Dust blowing around the cornstocks in field in southeast Colorado during Dust Bowl years. USDA photo.

temporarily eased the situation, and a dry spring and summer again intensified the shortages. By August extreme drought existed from northern West Virginia to southern New England. The Potomac River reached its lowest stream flow in history. New York and many other cities took stringent measures to curb water usage and to preserve existing water reserves. It was not until September and October 1966 that the drought was alleviated. Heavy rains, appropriately, fell on the day of the autumn equinox. Northeasters returned and the storms continued. By spring 1967 reservoirs returned to near-normal capacity, and the long drought finally ended.

Severe California Drought: 1976 and 1977 — California endured its driest year of record in 1977, after experiencing its third driest year of record the previous year. The combination of two dry years produced the most serious drought of this century in this most moisture-sensitive region. During the twenty-five months from November 1975 through January 1978, many stations received little more rainfall than would be expected during the average single rainfall season. Winter snowpack in the Sierras set new records for low snow accumulation, the flow of rivers and streams was reduced to record low amounts, and water-shortage reservoirs were drained to historic low levels.

The adverse conditions, however, were checked with surprising abruptness by heavy rainfall beginning in mid-December 1977. "Within six weeks the attention of many residents shifted from the problem of drought to the problem of flooding," Professor Marylyn L. Shelton of the University of California at Davis has pointed out.

During the most serious stage of the drought, nine million residents were placed on water-rationing programs because water supplies for over 100 communities were seriously depleted. Total losses of gross farm income attributed to the drought during 1977 were an estimated $2.4 billion, and losses would have been even

Opposite: Stump left high and dry when water level of Quabbin reservoir in west-central Massachusetts dropped during the Great Northeastern Drought in middle 1960s. Metropolitan District Commission, Massachusetts photo.

greater had it not been for the fact that 8,000 to 10,000 new irrigation wells were dug in the Central Valley to substitute groundwater for surface water. Approximately 7000 fires in brush and timberland burned 240,000 acres during 1977, and the fires cost an estimated $750 million. Reduced stream flow caused power plants to use fossil fuel at an expense of $650 million.

Longest Rainless Period in the United States — Bagdad, San Bernardino County, California, on Route 66 east of Barstow, had 993 consecutive days without measurable precipitation, from August 18, 1909, to May 6, 1912, and then another period of 767 consecutive days without measurable precipitation, from October 3, 1912, to November 8, 1914.

Least Annual Rainfall in the United States — Greenland Ranch, Inyo County, California, in Death Valley, had an average of 1.49 in (38 mm) annually for the 28-year period 1912 to 1940. More recently, the station at the National Park Headquarters in Death Valley, at an elevation of 194 ft (59 m) below sea level, has measured an average of 1.78 in (45 mm) annually.

September

By all these lovely tokens
September days are here,
With the summer's best of weather
And autumn's best of cheer.
— Helen Hunt Jackson, "September"

O wild West Wind, thou
breath of Autumn's bcing,
Thou, from whose unseen
presence the leaves dead
Are driven, like ghosts from an
enchanter fleeing.
— Percy Shelley, "Ode to the West Wind"

Up from the meadows rich with corn,
Clear in the cool September morn.
— John Greenleaf Whittier, "Barbara Frietchie"

Oh, shine on, shine on, Harvest moon, up in the sky." This is the month of greatest activity in gathering the fruits of the field and garden, and the task is aided for several days in a row by the presence of the best known and most widely celebrated full moon, adding its light to extend the evening hours available for harvest work. The full moon nearest the autumnal equinox is designated the Harvest Moon. The angle of the moon's orbit with the eastern horizon at this season is relatively small, and the times between daily risings are therefore shorter than in other months. Thus, the moon seems to be moving somewhat parallel to the horizon, so that at identical hours for several evenings it appears to be in the same position and lends its bright glow to light the harvest landscape.

Full moon is the time of syzygy, a fascinating sounding of syllables meaning that heavenly bodies are aligned in the same plane. This has meteorological significance, since sun and moon are then working in conjunction to raise a spring tide, and at the time of the equinoxes, tides produce their maximum levels. When these augmented tides occur in conjunction with a severe coastal storm or hurricane surge, great damage may result to coastal installations. The Great Miami Hurricane struck on September 18, 1926, and the New England Hurricane came on September 21, 1938, both accompanied by enormous syzygial tides. On the West Coast, the most destructive tropical storm of modern times, "El Cordonazo, the Lash of St. Francis," smashed the harbors of San Diego and Los Angeles on that saint's day, September 25, 1939, close to the time of syzygy.

September is a floating month on the meteorological calendar, moving back and forth from summer into autumn, and autumn into summer. Although at its start there maybe 90-degree temperatures in midafternoon, the early mornings of the closing days may be frosty. September's flower, the morning glory, has now grown tall and tangled; it opens later in the morning sunshine and closes earlier in the waning light of afternoon. For we are at the autumnal equinox, that point on the astronomical calendar when the sun rises in the true east and sets in the true west. The word *autumn* derives from the Latin *autumnum,* which Ovid called "the fairest season of the year."

In early September, the mean axis of the Pacific High attains its most northern position at 39°N, along the latitude of Punta Gorda and Cape Mendocino, the western points of Northern California, but its influence has already begun to wane with a weakening of its northeast arm, over British Columbia and Washington, a preliminary to the start of the winter storm season. The Aleutian Low, reappearing as a North American entity, has a mean center in the Bering Sea, close to Alaska's mainland, as it moves into a position to nourish autumn and winter storms.

In the Atlantic Ocean, the Azores–Bermuda High continues as an important weather traffic controller, though it has shifted slightly to the south and lost some of its summertime strength. A western extension is often found over the coastal waters from southern New England to Georgia. The Iceland–Greenland Low remains over the chill waters of Davis Strait, between Greenland and Iceland.

The Pacific jet stream makes its farthest-north entrance over Vancouver Island of British Columbia and pursues an almost due-east course close to the 49°N parallel, later crossing Lake Superior and the Atlantic Provinces of Canada.

Although September brings the first calendar days of autumn, its storm activity greatly resembles the summer conditions of August. Cyclonic storms are more active in Canada than in the United States, as demonstrated by the position of the primary continental storm track, close to 60°N in central Canada. In some years, storms from the northeast Pacific Ocean commence to enter western Canada, but this route will not become a regular feature of weather maps for another month.

Tropical disturbances cross the Atlantic Ocean from West Africa with greater frequency in September than in any other month. Some may recurve to the north in the Mid-Atlantic and move harmlessly over open waters east of Bermuda. Others continue on a steady western course and batter the islands of the West Indies before striking the American mainland. As in June, the western Caribbean and the southern Gulf of Mexico are breeding areas for tropical storms that can smash ashore at our Gulf Coast locations.

The principal tracks of high-pressure areas also resemble

those of August. A well-defined primary path extends across the northern United States near the 45°N parallel. An early portent of winter appears in the reactivation of high pressure in the Intermountain region of Idaho, Utah, and Nevada; a track leads east to a center of anticyclonic frequency of growing importance over West Virginia.

The southward retreat of the isotherms becomes rapid in September. Only south Florida and south Texas now show a plus 80°F (27°C) area. In the Southwest, the 90°F (32°C) line has disappeared and the 80°F (27°C) sector remains but is considerably contracted. The 70°F (21°C) isotherm runs from Chesapeake Bay west to southern Illinois and to southern Kansas, except for a dip south over the Appalachian highlands. The extreme North has dropped below 60°F (16°C), and islands of temperatures below 50°F (10°C) appear in the higher elevations of the Rocky Mountains.

Temperature extremes for September: maximum 126°F (52°C) at Mecca, California, on September 2, 1950; minimum −9°F (−23°C) at West Yellowstone, Montana, on September 24, 1926.

The greater part of the East now receives from 2 to 4 in (51 to 102 mm) precipitation except for some higher mountain areas where the catch is greater. The plus 4.0-in line encompasses only the immediate coastal plain from Cape Hatteras to south Texas, while the 8.0-in (203-mm) area is restricted to both coasts of south Florida. The rainy season in the North Pacific states usually begins sometime in September; this is reflected by the appearance of the 4.0-in line over the Olympic peninsula of northwest Washington, and the extension of the 2.0-in (51-mm) area south to the California border.

Farm activities are concentrated on the harvest of spring-planted crops and the preparation of soil for seeding winter grains. The latter begins in the northern section of the winter wheat belt during the first week of September, but the bulk of the crop elsewhere is not sown until after the tenth. Generous rains are needed at this time of year to condition the seed beds properly. In the South, cotton is being picked, and fortunately, the fall season averages the driest period in the Cotton Belt despite tropical storm visits in some years.

September

1 1955 Los Angeles hit 110°F (43°C) for all-time maximum record.

2 1935 Florida Keys Hurricane on Labor Day; small, intense center produced lowest U.S. pressure of 26.35 in (89.2 kPa); tide of 15 ft (4.6 m); 408 lost, many World War I veterans in CCC camp.

3 1821 Long Island Hurricane; landfall near site of present-day Kennedy Airport, then through western Connecticut; record tide at New York City; experience started William Redfield on his hurricane studies.

4 1939 "Once-in-a-Hundred-Years Rainstorm" at Washington, D.C.; 4.4 in (112 mm) in two hours.

5 1933 Hurricane hit Brownsville, Tex.; 40 killed; $12 million damage.

6 1881 Famous Yellow Day in Northeast, similar to Dark Day of 1780; caused by Michigan forest fires, where 20 villages burned; 500 people were killed; $2.3 million loss near Lake Huron.

7 1881 Hottest September day in Northeast: Washington, D.C., 104°F (40°C), Boston 102°F (39°C), New York City 101°F (38°C).
1888 Widespread killing frost in New England and New York; $1 million crop loss in Maine.

8 1900 The Galveston Disaster; severe hurricane tide inundated island city under 8 to 15 ft (2.5 to 4.5 m) of water; over 6000 perished; 3600 houses destroyed; damage $30 million.

9 1921 Most intense U.S. rainfall: 36.4 in (925 mm) fell in 18 hours, 38.2 in (970 mm) in 24 hours at Thrall, Tex.

10 1960 Hurricane Donna struck Florida Keys; barometer 27.46 in (93.0 kPa); gusts estimated at 180 mi/h (290 km/h); moved north over peninsula; Naples inundated; $300 million damage.
1980 Hurricane Allen came ashore above Brownsville, Tex.; dropped over 15 in (381 mm) rain near San Antonio; broke drought with heavy amounts through Rio Grande Valley.

11 1961 Hurricane Carla battered central Texas coast; 17.62-in (448-mm) rainfall; 45 deaths; $300 million damage.

12 1882 Hot, dry winds caused tree foliage to wither and crumble in east Kansas.
1979 Hurricane Frederic smashed into Mobile Bay area with 132-mi/h (213-km/h) sustained winds; $2.3 billion damage; costliest hurricane in American history.

13 1928 San Felipe Hurricane crossed Puerto Rico; "wind there was the highest, rainfall the heaviest, and destruction the greatest of record in recent years"; great damage in Virgin Islands; later hit Bahamas and Florida.

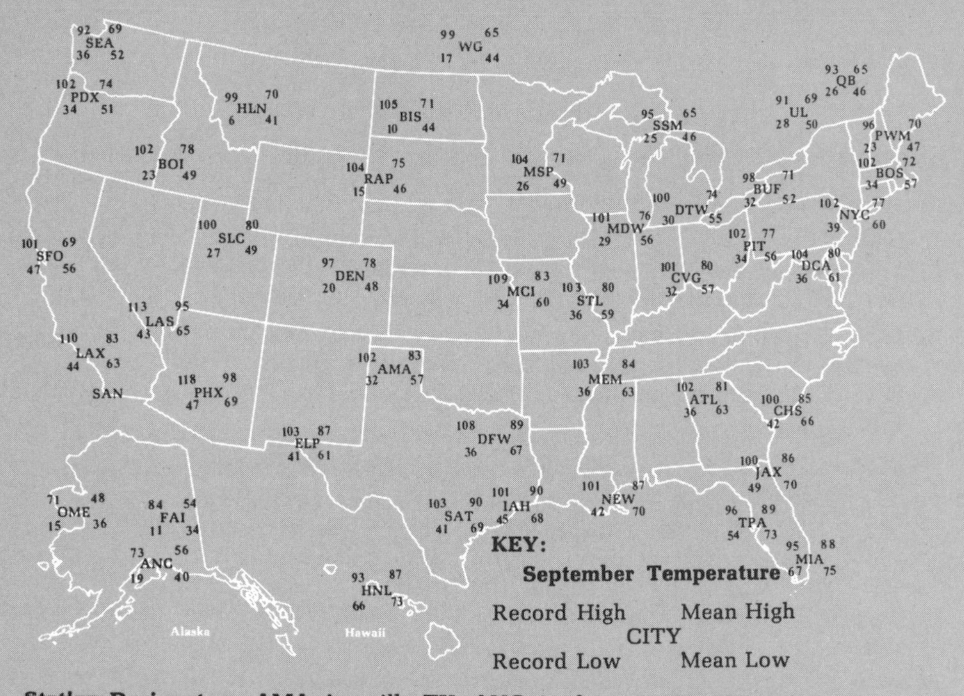

Station Designators: AMA Amarillo TX; **ANC** Anchorage AK; **ATL** Atlanta GA; **BIS** Bismarck ND; **BOI** Boise ID; **BOS** Boston MA; **BUF** Buffalo NY; **CHS** Charleston SC; **CVG** Cincinnati OH; **DCA** Washington DC; **DEN** Denver CO; **DFW** Dallas-Fort Worth TX; **DTW** Detroit MI; **ELP** El Paso TX; **FAI** Fairbanks AK; **HLN** Helena MT; **HNL** Honolulu HI; **IAH** Houston TX; **JAX** Jacksonville FL; **LAS** Las Vegas NV; **LAX** Los Angeles CA; **MCI** Kansas City MO; **MDW** Chicago IL; **MEM** Memphis TN; **MIA** Miami FL; **MSP** Minneapolis-St. Paul MN; **NEW** New Orleans LA; **NYC** New York NY; **OME** Nome AK; **PDX** Portland OR; **PHX** Phoenix AZ; **PIT** Pittsburgh PA; **PWN** Portland ME; **QB** Quebec QUE; **RAP** Rapid City SD; **SAN** San Diego CA; **SAT** San Antonio TX; **SEA** Seattle WA; **SFO** San Francisco CA; **SLC** Salt Lake City UT; **SSM** Sault Ste. Marie MI; **STL** St. Louis MO; **TPA** Tampa FL; **UL** Montreal QUE; **WG** Winnipeg MAN; **YC** Calgary ALB.

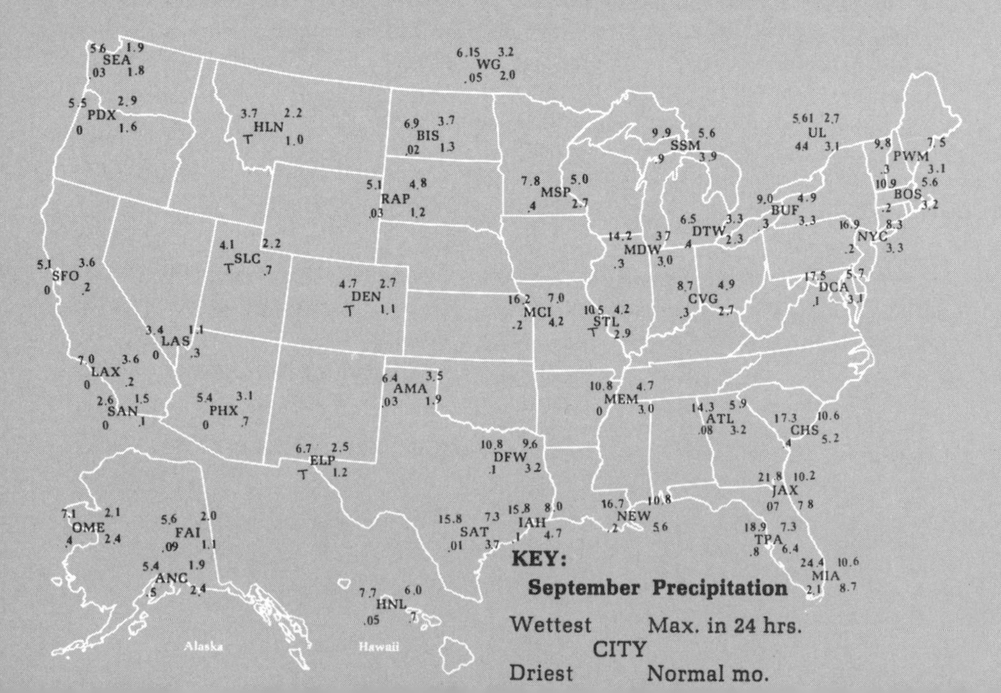

14 1944 Great Atlantic Hurricane swept Cape Hatteras with central pressure of 27.97 in (94.7 kPa); sideswiped New Jersey and Long Island, crossed southeast Massachusetts; 390 lost at sea.

15 1752 "The Great Hurricane" at Charleston, S.C.; tide within 12 in (30 cm) of covering downtown; water fell 5 ft (1.5 m) in 10 minutes with wind shift; tide height not equaled since.

16 1928 San Felipe Hurricane from Puerto Rico struck Palm Beach area; pressure of 27.43 in (92.9 kPa); enormous damage; floods at Lake Okeechobee drowned 1836 people, 1870 injured.

17 1963 Yuma's most intense rainfall since 1909: 2.42 in (61 mm) in 24 hours; exceeded by 2.72 in (69 mm) in August 1977.

18 1926 The Great Miami Hurricane; center passed over city with pressure at 27.61 in (93.5 kPa); wind 123 mi/h (198 km/h) for five minutes, 138 mi/h (222 km/h) for two minutes; tide 11.7 ft (3.6 m); 372 killed.

19 1947 Hurricane eye over New Orleans; 28.61 in (96.9 kPa); 51 killed; $110 million damage.

20 1845 Adirondack Tornado had 275-mi (443-km) track; waterspout across Lake Ontario, tornado over northern New York; spout on Lake Champlain; windfall in forest still discernible.

21 1938 Great Long Island–New England Hurricane; crossed Long Island (Bellport 27.94 in [94.6 kPa]); track west of New Haven, through Massachusetts and Vermont; massive forest blowdown; widespread floods; over 600 killed; damage in excess of $350 million.

22 1961 Hurricane Esther performed 350-mi (563-km) circle south of Cape Cod from September 21 to 25; then passed over Cape Cod and hit Maine.

23 1815 The Great September Gale; equal in stature to hurricane of 1938; track just to east with landfall near Moriches on Long Island and near Saybrook, Conn., then over Massachusetts and New Hampshire.

24 1926 Severe cold over Northwest; Yellowstone Park registered −9°F (−23°C), lowest ever in United States in September; severe freeze widespread, great crop destruction; earliest snow in Spokane Co., Wash.

25 1939 El Cordonazo tropical storm lashed Southern California; Los Angeles had 5.42 in (138 mm) in 24 hours; great floods; 45 lives lost; $2 million damage.

26 1950 Blue Sun and Moon over Northeast caused by forest fires in Alberta.

1963 San Diego's all-time maximum of 111°F (44°C); Los Angeles had near-record 109°F (43°C).

27 1816 Black frost over most of New England killed still unripened corn in north, causing a "scarce season."

1936 Early heavy snow of 21.3 in (54 cm) at Denver Airport in 60-hour storm.

28 1836 First of three heavy early 1836 snows: Hamilton, N.Y., 4 in (10 cm); Ashby, Mass., 2 in (5 cm).

1917 Pensacola Hurricane with wind gusts to 95 mi/h (153 km/h) and barometer at 28.50 (96.5 kPa); Mobile registered 75 mi/h (121 km/h).

29 1927 Tornado outbreak from Oklahoma to Indiana; major disaster at St. Louis, where 72 were killed; total deaths from outbreak, 81; damage $25 million.

30 1970 Fire disaster in interior San Diego County from September 25 to 30; strong Santa Ana winds consumed whole communities; 500,000 acres burned.

Hurricanes

Blow, winds, and crack your cheeks! rage! blow!
You cataracts and hurricanes, spout
Till you have drench'd our steeples, drown'd the cocks!
— William Shakespeare, *King Lear*, III, ii, 1

For I fear a hurricane;
Last night the moon had a golden ring
And tonight no moon we see.
— Henry Wadsworth Longfellow, "Wreck of the Hesperus"

Hurricanes represent the mature development of the tropical cyclone. Along with close relatives in the Pacific Ocean and Indian Ocean, they have been called "the greatest storms on earth." Fully developed hurricanes possess massive physical structures of great vertical extent reaching from the surface of the sea to the base of the stratosphere, cover enormous geographical expanses of land and sea, and have vigorous life spans continuing for many days. Great disasters, taking huge tolls in human lives and property, may result when a mature hurricane strikes a developed coastal area occupied by man and his creations.

A hurricane is a rotating wind system that originates over the warmth of tropical waters, possesses a central core having a higher temperature than the surrounding atmosphere, and is usually accompanied by torrential precipitation, occasional severe turbulence, and thunder and lightning. By definition, it must develop wind speeds of 74 mi/h (119 km/h); otherwise, the cyclonic whirl is designated either a tropical storm with winds ranging between 39 and 73 mi/h, or a lesser tropical depression if winds are 38 mi/h (61 km/h) or less. Tropical cyclones in their life cycles may pass through all three stages both when developing and when dissipating.

Structure — The hurricane is essentially a "heat engine," possessing a central core warmer than the surrounding atmosphere at respective levels. It derives its vast energies from the condensation of water vapor into water droplets that become visible as thick cloud. The attendant release of latent heat in this process provides the thermal fuel to energize the storm system, creating its vast wind force and precipitation potential.

There is an inward flow of air at lower levels below approximately 10,000 ft (3050 m); a middle layer has a predominately cyclonic circulation of rising air around the central core; and the upper third above approximately 25,000 ft (7600 m) exhibits an outward flow. At the center is a relatively calm area, known as the eye, around which walls of clouds circulate in vertical bands. Rising convective currents in these cloud structures create the extreme turbulence and excessive precipitation often accompanying the hurricane. The cloud system may extend upward in excess of 40,000 ft (12,200 m), with the highest cloud spreading out at the base of the stratosphere. The whole system has a progressive movement in accordance with the dominant steering current of the prevailing wind flow aloft.

Formation — Tropical storms require the proper combination of warmth, moisture, and instability of the atmosphere to initiate a circulation. The ocean waters must have a temperature of at least 80°F (27°C); the lower atmosphere must have both a high moisture content such as is usually present in low-pressure troughs at low latitudes, and a tendency to initiate rising currents of warm air that continue upward to considerable heights. The exact mechanism of tropical cyclone formation has been under intensive study in recent years, but the process is still not fully understood, since only about 10 percent of potential hurricane situations actually produce a storm circulation.

Most tropical storms are born in an "easterly wave" situation when a low-pressure trough in the upper atmosphere travels west at low latitudes, often moving from West Africa or adjacent waters across the entire North Atlantic Ocean to American shores. This disturbs the normally homogeneous tropical air mass, causing rising convective currents, cloud formation, and thunderstorm activity. Pressure falls over an area as a result of the ascending air columns. Around the disturbed area, the rotation of the earth north of 10°N provides a deflective effect, known as the Coriolis force, that diverts air flow into somewhat circular patterns. As convection increases and condensation and precipitation result, the release of more latent heat in this process adds more thermal energy to fuel the storm system. With continued fall of pressure, the cyclonically turning winds accelerate and a central eye forms, extending from the surface of the sea to great heights. A tropical storm with unknown potential and destination is born.

Eye — At the center of a well-developed hurricane lies the eye, a relatively calm area rising from the surface of the sea to the top of the storm structure, which is surrounded by walls of thick whirling cloud. Usually scattered cloud fragments are present, and sometimes sunlight may penetrate into the eye from above. The diameter of the eye varies greatly from storm to storm and from time to time in a single hurricane. In a mature stage the eye is usually a well-defined circular area, but in the declining stage it may become quite diffuse and spread out in an elliptical form. The mighty Palm Beach Hurricane in September 1928 had a central eye of 40 mi (64 km) breadth and a calm duration of forty minutes. The eye passage of a slow-moving center over Nassau, Bahama Islands, in September 1929, required four hours. Some minihurricanes have been observed with eye diameters of only 4 mi (6.4 km). The average diameter is about 15 mi (24 km), generally ranging from 12 to 30 mi (20 to 50 km, but sometimes ranging up to 60 mi (97 km). The eye has been clearly photographed by weather satellites from altitudes of 22,300 mi (35,888 km) above the earth.

Movement and Recurvature — The vast easterly wind current usually prevailing aloft over tropical latitudes in summer and autumn carries the disturbed area westward along the edge of the subtropical Azores–Bermuda high-pressure area. If this semipermanent anticyclone is strong and is in a normal position, the disturbance or "easterly wave" continues westward through the West Indies and into the Caribbean Sea or Gulf of Mexico. But if a trough of low pressure extends southward from temperate latitudes and the high pressure to the north weakens, the tropical disturbance may turn to the north and find a steering current associated with the trough. The path of the recurvature to the north depends on the structure and alignment of the trough. Some hurricanes veer to the northwest when near the West Indies and strike toward the coastline of the southeastern United States. Others continue the curving motion into the northeast, following a parabolic course over the open waters of the North Atlantic without striking land. The critical time for forecasting the future course of a hurricane occurs when the tropical controls of the easterly flow of low latitudes diminish and the westerly wind flow of temperate latitudes exerts a more vigorous influence.

Forward Speed — The New England Hurricane was moving at an extreme average speed of 58 mi/h (93 km/h) at landfall on central Long Island at 2:30 P.M. on September 21, 1938. In the tropics, however, the forward speed of a mature hurricane is usually from 10 to 15 mi/h (16 to 24 km/h), then becomes increasingly faster after recurvature to the north. But some have remained almost stationary for many hours or even days. Hurricane No. 3 in September 1878 gave Florida a four-day blow, remaining over land for three full days while crawling north over the peninsula at a 4-mi/h (6-km/h) pace.

Highest Wind — No anemometer has made a complete measurement of the peak winds of an extreme hurricane. Some instruments will stand such force, but either the supporting structure collapses or electric power fails. Observers have several times estimated the peak gusts at 200 mi/h (322 km/h) after their recorders failed. Engineers have calculated that winds of this force would be required to cause the extreme damage experienced, and theoretical studies confirm this estimate. Hurricane Inez produced a wind speed of 197 mi/h (317 km/h) on airborne instruments at an altitude of 8000 ft (2438 m) on September 28, 1966. The maximum winds in a hurricane occur in the wall cloud surrounding the central eye, and at the surface these will be found in the right semicircle of an advancing storm system.

The following readings of maximum gusts have been measured by anemometers near ground level:

175 mi/h (282 km/h), Hurricane Janet at Chetumal, Mexico, on September 27–28, 1955.

172 mi/h (277 km/h), Hurricane Camille at Main Pass Block, Louisiana, on August 17, 1969.

163 mi/h (262 km/h), Havana–Florida Hurricane at Havana, Cuba, on October 18, 1944.

161 mi/h (259 km/h), Hurricane Celia at Corpus Christi, Texas, on August 3, 1970.

160 mi/h (258 km/h), Fort Lauderdale Hurricane when at Hopetown, Bahama Islands, on September 16, 1947.

Lowest Pressure — A mercurial barometer reading of 26.185 in (88.67 kPa) was observed at sea on SS *Sapoerea* when 460 mi (740 km) east of Luzon, Philippine Islands, on August 18, 1927. A low of 26.35 (89.23 kPa) was calculated from tests on an aneroid barometer exposed on land at Matecumbe Key, Florida, during the Labor Day Storm on September 2, 1935.

Aircraft reconnaissance planes have dropped sounding instruments (dropsondes) into the eye of hurricanes and typhoons and have calculated some extremely low barometric readings. A new world record of 25.69 in (87.0 kPa) was obtained in Super Typhoon Tip on October 12, 1979, when about halfway between central Luzon Island and Iwo Jima. In the Western Hemisphere, Hurricane Camille produced a reading calculated from an aircraft sounding of 26.73 in (90.5 kPa) on August 17, 1969, over the north-central Gulf of Mexico prior to its landfall that night on the Mississippi coast.

Opposite: Two unidentified yachtsmen cling to a tree as they watch their boat smash against the rocks at Wollaston Beach, Quincy Bay, near Boston, August 31, 1954. Photo by Charles W. Flagg.

Storm Surge — As a hurricane crosses the continental shelf and moves onto the shoreline, water levels can increase by 5 to 10 ft and sometimes by more than 15 ft (1.5 to 4.6 m). The surge occurs slightly in advance and to the right of the path of the center of the eye of the storm. The raised water level is imposed on normal tides, and in turn, wind-driven waves are superimposed on the surge. The build-up of water levels can cause severe flooding on islands and in coastal areas, particularly when the storm surge coincides with local high tides. Because much of the immediate coastline of the Atlantic and Gulf lies less than 10 ft (3 m) above mean sea level, the danger from a storm surge is great. The loss of life and wind damage to property escalates many times as a result of the coastal and inland flooding from a storm surge. The real disaster at Galveston in September 1900 was caused by the rush of a 5-ft (1.5-m) quick rise on top of a 15-ft (4.6-m) tide flooding on the island. Downtown Providence, Rhode Island, was severely inundated in 1938 and 1954 by storm surges funneling up a narrowing river estuary. The storm surge is one of nature's greatest killers in the Far East; a Bay of Bengal cyclone in East Pakistan drowned 200,000 persons in 1970 according to official estimates, though other figures ran as high as 500,000. Evacuation of low-lying areas is essential long before the arrival of a hurricane and its storm surge.

Precipitation — The greatest rainfalls in the United States have resulted from tropical storms. Claudette on July 24–25, 1979, precipitated a deluge of 43.0 in (1092 mm) near Alvin, Texas, in twenty-four hours. Other heavy falls were 38.7 in (983 mm) at Yankeetown, Florida, on September 5–6, 1950, and 38.2 in (970 mm) at Thrall, Texas, on September 9–10, 1921. In the North, Camille brought an estimated 27 in (686 mm) to Massies Mill, Virginia, on August 19–20, 1969; Diane dropped 19.4 in (493 mm) at a weather station at Westfield, Massachusetts, on August 18–19, 1955; and Agnes let loose 18.8 in (478 mm) over Pennsylvania on June 21–23, 1972, on an agricultural research site in western Schuylkill County. Most hurricanes are attended by more moderate rainfalls in the range of 1 to 5 in (25 to 127 mm), and these amounts form an important portion of the annual water budget in tropical areas.

A serious drought in Texas was ended in the summer of 1980 by the combined rains of Hurricane Allen and Tropical Storm Danielle.

Season — June storms are usually small and of minor intensity, though Audrey's striking in Louisiana on June 27, 1957, was a tragic exception. The Gulf of Mexico and the western Caribbean Sea spawn many of the early-season storms affecting the United States. Activity increases slightly in July and storms grow in size, but major hurricanes are infrequent. A marked increase takes place in August in both frequency and intensity, and the scene of major activity shifts eastward.

The height of the hurricane season is reached in early September, when the tropical zone of the broad Atlantic Ocean becomes the principal breeding place. A third of all North Atlantic storms occurs in this month, and some giant ones cross the entire expanse from West Africa to North America. There is a slight decrease in the second half of the month, then a marked increase during the first two weeks of October. At this time the Caribbean Sea again becomes an active source region. The season goes into a steady decline in the second half of October, and few storms are charted in November in either the Atlantic Ocean or the Caribbean Sea.

Most Active Season — The year 1933 produced twenty-one tropical cyclones from May 4 to November 16; nine were full hurricanes. There were eighteen tropical cyclones in 1969, and seventeen in 1887. The most hurricanes for a season — twelve — were produced in 1969, and both 1916 and 1950 had eleven full hurricanes.

Least Active Season — In 1914 no hurricanes and only one tropical storm appeared; the latter moved through southern Georgia and along the Gulf Coast. Only one hurricane was mapped in 1890, and it stayed well away from the American coast; no tropical storms occurred that season. The only other year without a single hurricane was 1907, but four tropical storms developed.

Most Activity in a Single Month — Eight tropical cyclones were charted in action during some part of August 1933, and seven in September 1933 and August 1936, and six in September 1964. Four full hurricanes co-existed on August 22, 1893, but no single date since 1900 has had four. Three full-fledged hurricanes were mapped on September 11, 1961: Carla, Debbie, and Esther; also three on September 14–15, 1967: Beulah, Chloe, and Doria. But four tropical storms were in existence simultaneously on October 17, 1950.

Earliest — Circulations of hurricane force in advance of the normal season were reported in May 1889, March 1908, and May 1951 in the Windward Islands area, but their careers were short-lived. About once every three years a tropical storm of usually minor intensity will develop in the November to May period. The earliest crossing of the United States coastline by a full-fledged hurricane was Alma's arrival in northwest Florida on June 9, 1966. The earliest by a lesser tropical cyclone came in northern Florida on May 24–25, 1970.

Latest — A tropical storm developed in the western Caribbean Sea at the end of November 1925. After crossing central Florida on the thirtieth, it developed hurricane strength over Atlantic waters for a few hours but was down to tropical storm intensity when making a second landfall near Cape Hatteras on December 2. A onetime hurricane, roaming the western Atlantic from November 17 to December 2, 1888, sideswiped Cape Cod on November 27 with gale-force winds, causing considerable damage. More recently, Hurricane Alice moved southeast through the Leeward Islands from December 30, 1954, to January 5, 1955, qualifying as the earliest or the latest hurricane of record, according to one's predilections. A maverick tropical storm crossed South Florida on February 2, 1952.

Longest and Shortest Lifetime — Tropical cyclone Ginger wandered about the west-central North Atlantic, the Bermuda Triangle, and the coasts of North Carolina and Virginia for 31 days, from September 5 to October 5, 1971. On 20 of these days it was classified as a hurricane, on the remainder a tropical storm. There have been many tropical cyclones that attained hurricane intensity for periods of only 12 hours or less.

Erratic Courses — Hurricanes often take unexpected paths. The famous Yankee Hurricane reached Miami at a late date and from an unexpected direction. It formed near Bermuda in late October 1935, moved west toward Cape Hatteras, but then turned southwest, and passed right over Miami before entering the Gulf of Mexico, where it performed a loop and headed toward the peninsula again. Hurricane Ginger was born just east of the Bahamas in early September 1971 and moved northeast to 50°W, well east of Bermuda, then reversed to a south-southwest course that took it back near its birthplace. A northwest course to the North Carolina coast followed and another reversal carried it eastward about 400 mi (644 km) into the Atlantic, where it dissipated on October 5, after a month's journey.

Great Hurricanes — Each century has produced one or more candidates for the title of "greatest hurricane" since Christopher Columbus first witnessed one. He experienced the fringes of a hurricane or tropical storm in mid-September 1494 when near Hispaniola and felt the full force of another near Santo Domingo on July 10, 1502.

The "Great Hurricane of October 1780" at Barbados has long been notorious for its almost complete devastation of the land and for its influence in altering the naval balance between France and Great Britain in the War of American Independence. In the nineteenth century the "Great Hurricane of October 1846" struck devastation through Cuba, at Key West, and along the entire Atlantic seaboard; it was outstanding for size, duration, and destructive power.

In the present century, San Felipe in 1928 is thought to have been the most violent and destructive. After causing great damage on Guadeloupe, St. Kitts, and Montserrat, it smashed across the width of Puerto Rico on September 13, 1928; during the passage, 300 were killed and damage of $50 million resulted. San Felipe made another landfall at Palm Beach, Florida, on September 16, and 1836 persons died of drowning in the Lake Okeechobee area and damage to the extent of $25 million resulted throughout south Florida. Additional losses marked its path north along the Atlantic seaboard.

In the more recent past, Camille in August 1969, with its double punch of wind and tide in Mississippi and flooding deluges in Virginia, proved the most costly, in terms of dollars, up to that time and is judged to have possessed the most concentrated destructive power of any hurricane to make a landfall on a built-up portion of the United States mainland. Its central pressure upon landfall in Mississippi was 26.84 in (90.9 kPa).

Most Deadly — City officials estimated that over 6000 persons of a population of 38,000 were missing at Galveston, Texas, after the flood waters had subsided from the "great storm" on September 8–9, 1900. Recent estimates for the entire area have ranged up to 7200 — the exact number can never be ascertained. A storm wave, said to be "nearly five feet high," struck the already inundated waterfront early in the evening, causing houses and buildings to either collapse or float away. Many who thought they had reached a safe vantage in these structures were drowned.

The Louisiana bayou country was submerged by a storm wave surging from the Gulf of Mexico on October 1–2, 1893, and 2000 people died. The South Carolina coast was battered earlier that year, on August 27–28, with losses estimated between 1000 and 2000. A storm-induced tide on Lake Okeechobee on September 16–17, 1928, previously mentioned, broke dikes and drowned 1836 Floridians.

In recent years the big killers were Audrey in Louisiana on June 27, 1957, with 390 listed fatalities, and Camille on August 17–20, 1969, with 256 victims, mainly in Mississippi and Virginia. Flora, in 1963, took a toll of about 6750 dead in Haiti and Cuba.

Most Costly — The potential for destruction has increased greatly in recent years as a result of the concentrated development of coastal resorts, and rampant inflation has skyrocketed the dollar figures. The 1979 season was the most recent to experience a mighty hurricane making a landfall on a built-up area. Hurricane Frederic smashed ashore near the entrance to Mobile Bay on September 12, 1979, and caused damage placed at $2.3 billion, slightly exceeding the $2.1 billion damages assigned to the flood water losses during Hurricane Agnes's deluges in Pennsylvania and New York. In the previous decade Camille was the costliest at $1.4 billion, resulting from its landfall on the Mississippi and Louisiana coasts on August 17–18, 1969, and its flood waters in Virginia two days later. The average annual loss over the period 1972–76 was $771 million. But with Agnes excluded, the average for the period 1973–76 reduces to $189.5 million. The figure within the United States in 1973 was only $18 million, mostly crop damage from heavy rains, since no urban area was struck.

Hurricane Disaster-Potential Scale

(Developed by Herbert Saffir, Dade County, Florida, consulting engineer, and Dr. Robert H. Simpson, former National Hurricane Center director)

Category	*Central Pressure* (kiloPascals)	(inches)	*Winds* (mi/h)	*Surge* (feet)	*Damage*
1.	98.0 or more	28.94	74–95	4–5	Minimal
2.	96.5–97.9	28.50–28.91	96–110	6–8	Moderate
3.	94.5–96.4	27.91–28.47	111–130	9–12	Extensive
4.	92.0–94.4	27.17–27.88	131–155	13–18	Extreme
5.	91.9 or less	27.16 or less	156 or more	18.1 or more	Catastrophic

The Saffir/Simpson Damage-Potential Scale

The Saffir/Simpson Damage-Potential Scale is used by the National Weather Service to give public-safety officials a continuing assessment of the potential for wind and storm-surge damage from a hurricane in progress. Scale numbers are made available to public-safety officials when a hurricane is within 72 hours of landfall.

Scale numbers range from 1 to 5. Scale No. 1 begins with hurricanes in which the maximum sustained winds are at least 74 mi/h (119 km/h) or which will produce a storm surge 4 to 5 feet (1.4 m) above normal water level. Scale No. 5 applies to those in which the maximum sustained winds are 156 mi/h (251 km/h) or more, or which have the potential of producing a storm surge more than 18 feet (5.5 m) above normal. Atmospheric-pressure ranges have been adapted to this scale, and pressure ranges associated with each are listed above.

The scale numbers are not forecasts but are based on observed conditions at a given time in a hurricane's life span. They represent an estimate of what the storm would do to a coastal area if it were to strike without change in size or strength. Scale assessments are revised regularly as new observations are made, and public-safety organizations are kept informed of the hurricane's disaster potential.

The damage-potential scale indicates probable property damage and evacuation recommendations as listed below:

1. Winds of 74 to 95 mi/h (119 to 153 ki/h). Damage primarily to shrubbery, trees, foliage, and unanchored mobile homes. No real damage to other structures. Some damage to poorly constructed signs. And/or storm surge 4 to 5 feet (≃1.5 m) above normal. Low-lying coastal roads inundated, minor pier damage, some small craft torn from moorings in exposed anchorage.
2. Winds of 96 to 110 mi/h (154 to 178 km/h). Considerable damage to shrubbery and tree foliage; some trees blown down. Major damage to exposed mobile homes. Extensive damage to poorly constructed signs. Some damage to roofing materials of buildings; some window and door damage. No major damage to buildings. And/or storm surge 6 to 8 feet (≃2 to 2.5 m) above normal. Coastal roads and low-lying ecape routes inland cut by rising water 2 to 4 hours before arrival of hurricane center. Considerable damage to piers; marinas flooded. Small craft torn from moorings in unprotected anchorages. Evacuation of some shoreline residences and low-lying island areas required.
3. Winds of 111 to 130 mi/h (179 to 209 km/h). Foliage torn from trees; large trees blown down. Practically all poorly constructed signs blown down. Some damage to roofing materials of buildings; some window and door damage. Some structural damage to small buildings. Mobile homes destroyed. And/or storm surge 9 to 12 feet (≃2.6 to 3.9 m) above normal. Serious flooding at coast; many smaller structures near coast destroyed; larger structures near coast damaged by battering waves and floating debris. Low-lying escape routes inland cut by rising water 3 to 5 hours before hurricane center arrives. Flat terrain 5 feet (1.5 m) or less above sea level flooded inland 8 miles (≃13 km) or more. Evacuation of low-lying residences within several blocks of shoreline possibly required.

4. Winds of 131 to 155 mi/h (211 to 249 km/h). Shrubs and trees blown down; all signs down. Extensive damage to roofing materials, windows, and doors. Complete failure of roofs on many small residences. Complete destruction of mobile homes. And/or storm surge 13 to 18 feet (~4 to 5.5 m) above normal. Flat terrain 10 feet (~3 m) or less above sea level flooded inland as far as 6 miles (~10 km). Major damage to lower floors of structures near shore due to flooding and battering by waves and floating debris. Low-lying escape routes inland cut by rising water 3 to 5 hours before hurricane center arrives. Major erosion of beaches. Massive evacuation of all residences within 500 yards (~455 m) of shore possibly required and evacuation of single-story residences on low ground within 2 miles (~3 km) of shore required.
5. Winds greater than 155 mi/h (249 km/h). Shrubs and trees blown down; considerable damage to roofs of buildings; all signs down. Very severe and extensive damage to windows and doors. Complete failure of roofs on many residences and industrial buildings; extensive shattering of glass in windows and doors. Some complete building failures. Small buildings overturned or blown away. Complete destruction of mobile homes. And/or storm surge greater than 18 feet (~5.5 m) above normal. Major damage to lower floors of all structures less than 15 feet (~4.5 m) above sea level within 500 yards (~455 m) of shore. Low-lying escape routes inland cut by rising water 3 to 5 hours before hurricane center arrives. Massive evacuation of residential areas on low ground within 5 to 10 miles (~8 to 16 km) of shore possibly required.

Most Intense in the United States

Hurricane	*Category*	*KiloPascals*	*Inches*
Florida Keys, 1935	5	89.2	26.35
Camille, 1969	5	90.9	26.84
Florida Keys and Texas, 1919	4	92.7	27.37
Okeechobee, 1928	4	92.9	27.43
Donna, 1960	4	93.0	27.46
Texas (Galveston), 1900	4	93.1	27.49
Louisiana (Grand Isle), 1909	4	93.1	27.49
Louisiana (New Orleans), 1915	4	93.1	27.49
Carla (Texas), 1961	4	93.1	27.49
Florida (Miami), 1926	4	93.5	27.61

Deadliest in the United States

Hurricane	*Number of victims*
Galveston, 1900	6000–7000
Louisiana, 1893	2000
South Carolina, 1893	1000–2000
Okeechobee, 1928	1836
Florida Keys and Texas, 1919	600–900*
Georgia and South Carolina, 1881	700
New England, 1938	600
Florida Keys, 1935	408
Audrey, Louisiana, 1957	390
Atlantic Coast, 1944	390*

Costliest in the United States

Hurricane	*Millions of dollars in contemporary value*
Frederic, 1979	$2,300.
Agnes, 1972	2,100.
Camille, 1969	1,420.7
Betsy, 1965	1,420.5
Diane, 1955	831.7
Eloise, 1975	490.
Carol, 1954	461.
Celia, 1970	453.
Claudette, 1979	400.
Carla, 1961	408.
Donna, 1960	387.
David, 1979	320.
New England, 1938	306.

Tropical Storms and Hurricanes of the West Coast of Mexico

The warm waters of the eastern Pacific Ocean between 12°N and 25°N provide a spawning ground for tropical storms in the warm season from June to November. The area of activity extends from the Mexican and Central American coast westward to the waters south of the Hawaiian Islands. On an average, fifteen storms generate each year, with seven achieving full hurricane stature. Although they are generally smaller in size than their Atlantic

* Principally on ships at sea.

cousins, they can attain equal severity over a restricted area. The most intense storms usually generate early or late in the season, forming close to the coast and quite far south. August and September are the most frequent breeding months, with an average of 4.5 and 4.1 storms respectively.

The normal storm track trends toward the west or northwest through an ocean area little traversed by commercial vessels other than the tuna fleet. Occasionally one may recurve to the north and northeast and strike a blow at the west coast of Mexico and Baja California. The forward speed of movement is from 7.5 to 18 mi/h (12 to 29 km/h). The average observed duration lasts from four to five days, though several exceptional ones have been tracked for over ten days. Most storms move to the northwest and reach the cooler waters of the central Pacific Ocean, well offshore, where they dissipate rapidly. But some have carried moist airstreams across northwest Mexico into the border states of the desert Southwest where they have caused damaging floods and mudslides.

While the radius of gale-force winds is relatively small, the storms are capable of raising winds of 100 to 140 mi/h (161 to 225 km/h). Extreme destruction has occurred at several of the principal Mexican ports when full-fledged storms strike. The heavy rains attending most of them are very beneficial to agriculture in this usually very dry area.

These tropical storms have been known to range as far north as Southern California and as far west as the Hawaiian Islands. In the very active month of September 1939, no fewer than five tropical storms moved northwest along the coast of Baja California; two came ashore below Ensenada, within 200 mi (322 km) of the United States border, and another reached 32°N before turning east to come ashore between San Diego and Los Angeles.

Southern California received its worst battering from this storm on September 24–25, 1939, when forty-five lives were lost at sea and damage of approximately $2 million was suffered on land. Winds reached a sustained force of 32 mi/h (52 km/h) in downtown Los Angeles, inland from the sea. Record September rains fell, with the Civic Center catching a 24-hour total of 5.42 in (138 mm), and the gauge at Mt. Wilson Observatory (elevation 5850 ft [1783 m]) gathered 11.6 in (295 mm).

Not for another thirty-seven years did another strong tropical

storm of whole-gale intensity penetrate into the United States, though the dying circulations of several have brought copious moisture northward into the deserts of the Southwest. On September 9–10, 1976, Hurricane Kathleen moved over Baja California, entered the United States close to Yuma, Arizona, and then tracked north into the Nevada desert. It gave Yuma a wind lashing with a measured mile at 57 mi/h (92 km/h). Very heavy rains caused substantial damage to highways and bridges, and mudslides resulted in destruction of some buildings. The village of Ocotillo near Palm Springs, California, was inundated by flood waters and six lives were lost. Excessive rains of 8.00 in (203 mm) were measured at Mt. San Jacinto, in Riverside County, and 7.28 in (185 mm) at Big Pines, in Los Angeles County. Heavy rains also fell in western Arizona, where Kofa Mountain, Yuma County, picked up 5.42 in (138 mm). Rains covered much of the Nevada desert as well.

Since 1950, five storms with hurricane-force winds have passed near enough to the Hawaiian Islands to inflict damage on some part of the state. These all generated in the ocean to the east or south. In August 1950, September 1957, and December 1957, storms brushed the islands and caused high surf. On August 7, 1959, Hurricane Dot moved directly across Kauai with a high wind report of 100 mi/h (161 km/h). Total damage was estimated at $6 million. Hurricane Fico on July 20, 1978, passing within about 150 mi (241 km) of the Island of Hawaii, gave the south shore a wind lashing, and raised heavy seas.

A meteorological oddity occurred in 1971 when Hurricane Irene from the Caribbean Sea crossed the mainland of Central America through Nicaragua and emerged into the Pacific Ocean on September 20. It was renamed Oliva when it quickly regenerated into a full-fledged hurricane. The feat was repeated in September 1974 by Atlantic Fifi's becoming Pacific Orlene and in September 1978 by Atlantic Greta's becoming Pacific Olivia.

Opposite: Destruction scene on the Naugatuck River at American Brass Company, Waterbury, Connecticut, resulting from Hurricanes Connie and Diane, August 1955. Courtesy of The Anaconda Company, Brass Division.

Four Great Hurricanes

The four hurricanes discussed below represent the most destructive ever to have battered their respective coastlines. The Galveston Hurricane and Tide in September 1900 took by far the greatest toll of lives of any storm in the United States. The Florida Keys Hurricane in September 1935 was the most intense in regard to tight-knit structure and wind power ever to strike Florida. The Long Island and New England Hurricane in September 1938 demonstrated that a full-fledged hurricane can cause enormous damage and tragedy even in distant northern latitudes. It remained for Hurricane Camille in August 1969 to build up the greatest combination of wind force and wave surge ever to strike a well-populated coast of the United States.

The Great Hurricane Wave at Galveston in September 1900 — The wind maintained a constant northerly direction through the night and the tide level continued to rise steadily, facts noticed by Galveston's weatherman Isaac Cline when he visited the beach soon after dawn on the fateful Saturday, September 8, 1900. Because of his acute understanding of the significance of these concurrent conditions, several thousand lives were saved later that day and night. Yet the feeling seemingly inbred in man that "it can't happen here" caused the death before midnight of over six thousand souls who might otherwise have lived long full lives, and Cline's underestimation of the total force of nature's coming onslaught led to his own grievous personal loss and a harrowing experience that evening.

Long ocean swells rolling in from the southeast had been noticed the previous day along the popular beaches of narrow Galveston Island, facing the sea. The surf was considered too rough by late afternoon for the usual last dip before supper in the refreshing waters.

The long rollers were the harbingers of a tropical disturbance known to be churning about somewhere in the vast reaches of the Gulf of Mexico. Warnings of its presence had been distributed two days earlier when a storm system was reported moving westward through the Straits of Florida near Key West and entering the wide Gulf, pursuing an apparent course to the west-northwest.

Lacking the instant communication of radio from air reconnaissance and the electronic seeing-eye of radar, forecasters could only surmise the exact location and true path of the storm center. Nor did they have any information concerning its strength — whether a tropical storm of moderate proportions or a full-fledged hurricane — to give a clue as to its potential destructive force.

Cline's anxiety was raised further by the situation at his early morning check of the meteorological instruments. Despite the barometer's dropping only a tenth of an inch overnight and the presence of patches of blue in the early morning sky, he had an ominous feeling. Accordingly, he filed a telegram to the central forecast office in Washington: *Unusually heavy swells from the southeast, intervals from one to five minutes, overflowing low places south portion of city three to four blocks from beach. Such high water with opposing winds never observed previously.*

The wave crests were increasing in magnitude and frequency during the morning. With a continued wind flow from the north, Cline reasoned that a storm wave of dangerous proportions would build up. The direction of the wind indicated that the storm's track must be considerably south of the usual path, and this would place Galveston Island in the northwest quarter of the approaching center. If it passed to the south, the low-lying sand spit would be in the dangerous right semicircle of the approaching storm, where wind flow from seaward and the strong storm surge might combine to create an overwhelming inundation of the low island, which rose only 8.7 ft (2.7 m) above sea level at its highest point. Somewhat similar threats had arisen during hurricanes in 1837, 1867, and 1875, but the attending circumstances did not combine to produce a complete inundation on those occasions.

In the absence of any civil defense organization or community awareness program, Cline took it upon himself to play the role of a meteorological Cassandra and rode up and down the beaches on horseback warning people of the impending danger. He urged residents near the beaches to move to higher ground near the center of the city and advised summer visitors still in Galveston to take the first train for Houston.

All the while, the tide continued to rise, the winds to mount higher and higher, the sky to cloud over completely, and sporadic bursts of rain to fall. Soon the sea invaded all of the southern part of the island, while the north wind drove the pent-up waters of the bay higher and higher over the back part of the land. Sometime close to four o'clock in the afternoon the surging tides met, and the water rose to the depth of a foot even at the highest point of land along Broadway. Soon the waters entered the lobby of the Tremont Hotel and stood thigh-deep, forcing hundreds of refugees to the upper floors.

At 6:15 P.M., the Weather Bureau's anemometer was ripped from its supports after hitting a steady rate of 84 mi/h (135 km/h) with gusts to 100 mi/h (160 km/h), and then the instrument shelter atop the roof, along with its thermometers and psychrometers, took off on an aerial voyage. Remaining on duty, Observer Blagden saw his barometer drop below 29.00 in (98.2 kPa), indicating a storm of class *3* or *4* in the making. It would eventually plummet to an estimated 27.91 in (94.5 kPa) when the eye passed to the south of the island between 8:00 and 9:00 P.M.

As darkness settled over the island, battered by wind and beleagured by rising water, the crescendo of the wind continued as it shifted gradually from northeast to east, at speeds estimated at 120 mi/h (193 km/h); and the barometer slipped steadily downward to reach its minimum at about 8:30 P.M. The buildings still standing were now beset by twin destructive forces. The surge of the ever-rising storm tide, topped by gigantic waves, was undermining foundations and even lifting entire structures from their blocks, while at the same time the upper floors were being battered by the hurricane winds and torn asunder, their fragments becoming airborne missiles.

The horrors experienced by thousands of Galvestonians that dreadful evening and night can best be portrayed in the sad words of Isaac Cline:

> By 8 p.m. a number of houses had drifted up and lodged to the east and southeast of my residence, and these with the force of the waves acted as a battering ram against which it was impossible for any building to stand for any length of time, and at 8:30 p.m., my residence went down with about fifty persons who had sought it for safety, and all but eighteen were hurled into eternity. Among the lost was my wife, who never rose above the water after the wreck of the building. I was nearly drowned and became uncon-

scious, but recovered through being crushed by timbers and found myself clinging to my youngest child, who had gone down with myself and wife. My brother joined me five minutes later with my other two children, and with them and a woman and child we picked up from the raging waters, we drifted for three hours, landing 300 yards from where we started. There were two hours that we did not see a house nor any person, and from the swell we inferred that we were drifting to sea, which, in view of the northeast wind then blowing, was more than probable. During the last hour that we were drifting, which was with southeast and south winds, the wreckage on which we were floating knocked several residences to pieces. When we landed about 11:30 p.m., by climbing over floating debris to a residence on Twenty-eighth Street and Avenue P, the water had fallen 4 feet.

The Toll — Herbert Molloy Mason, Jr., has summarized the destruction in *Death from the Sea* (1972), the best and most recent retelling of the story of the disaster:

> In Galveston itself, the area of total destruction exceeded 1500 acres, including the denuded eastern end and a stretch of beachfront extending for more than 5000 yards. Here, 2636 houses were knocked to pieces, and elsewhere in town and further down the island another 1000 homes were destroyed, bringing the total to 3636. And the dead. Approximately 3000 corpses were pulled from the wreckage or buried among the debris in the demolished areas east and along the beachfront. Another thousand were strewn in the streets and the yards; 500 were gathered on the shores of the west and north bay areas, and it was guessed that something like 500 were swept out to sea during the furious hours when the mountains of water crashed against the town. West of town, along the slender finger of sand pointing westward for almost thirty miles, an estimated 1200 were killed out of a total population of 1600. The total killed, then, on Galveston Island reached more than 6000, or nearly 18 percent of the population. Estimates of property damage finally reached $30 million . . . Missing and known dead outside of Galveston were estimated to number a thousand. The precise casualty figure has never been arrived at, and there is hardly any way it could be; but any way the figures are totaled the sum does not drop below 7200 dead.

The Florida Keys Hurricane on Labor Day in 1935 — The most concentrated wind-and-wave fury ever to strike the mainland of the United States smashed across the low-lying central Florida

Keys on Monday evening, September 2, 1935, taking a toll of over 400 lives. It was Labor Day afternoon when the storm struck, and the coincidence of its occurrence on a holiday was responsible for much of the high death count. An emergency rescue train, ordered to proceed from the mainland down the string of keys and return with a group of World War I veterans then living in C.C.C. camps, mainly on Matecumbe Key, experienced administrative delays in making up on a holiday afternoon; and more valuable time was lost as a result of a mishap along the tracks when the train was nearing the threatened area. If the rescue train had arrived and departed two hours earlier, as scheduled, much of the human tragedy that night might have been prevented.

There were plenty of warnings. The Jacksonville Hurricane Forecast Office early on the morning of the second hoisted northeast storm warnings from Fort Pierce on the east coast around to Fort Myers on the west coast. At 3:30 A.M. on the fateful Monday, the following advisory was issued:

> Tropical disturbance still of small diameter but considerable intensity is moving slowly westward off the coast of west-central Cuba, attended by shifting gales and probable winds of hurricane force over a small area. It will probably pass through the Florida Straits Monday. Caution is advised against high tides and gales on the Florida Keys and for ships in its path.

A disturbed area was first observed on August 29 to the east and north of Turks Island in the Bahamas. It soon concentrated around a central eye and displayed increasing energy. Hurricane intensity was reached on September 1 when near the south end of Andros Island about 220 mi (354 km) southeast of Miami; the storm center was heading westward, with a slightly southward curvature, carrying over the Great Bahama Bank and into the Florida Straits. Without the modern probes of aerial reconnaissance and searching radar, no exact knowledge of the hurricane's true location, movement, and intensity was available.

On Monday, September 2, the center of the now violent hurricane veered to the west-northwest toward the central Keys, at the same time coiling into a very tight whirl of wind and wave. While at peak intensity in the evening, the class 5 hurricane passed over Long Key and Matecumbe Key, which lie a little less than halfway down the 90-mile chain of small coral islands stretching from Key Largo on the east to Key West. An aneroid barometer on Long Key, later tested for accuracy, dropped to the lowest reading ever observed within the United States, 26.35 in (89.2 kPa), and other instruments indicated readings below 27.00 in (91.4 kPa).

All wind instruments were destroyed, but from an analysis of the damage the gusts were estimated at 150 to 200 mi/h (241 to 321 km/h) over an area of about 15-mi (24-km) radius on the eastern side of the eye of the storm. Unusually high flooding of the keys took place during the approach of the storm, but with the arrival of the hurricane's eye a wall of water estimated to have been 20 ft (6 m) high completely inundated the two keys. Wave wash marks on trees were as high as 30 ft (9 m) above mean sea level.

The wind lulled briefly between 8:00 and 9:00 P.M. at Alligator Reef, with the direction chopping sharply from northeast to southeast; this point was on the northern border of the eye. The calm on Long Key lasted for 55 minutes, from 9:20 to 10:15 P.M., with the wind coming finally in a blast out of the south-southwest. On Lower Matecumbe Key the calm continued for 40 minutes. With the rate of movement estimated at 10 mi/h (16 km/h), the diameter of the eye was calculated to be about 8 mi (13 km).

The area of complete destruction covered about 30 mi (48 km) in width across the narrow islands, from the vicinity of Tavernier on the east to Vaca Keys on the west. In this area all structures were either swept from their locations or badly damaged by the battering winds and surging seas. The tracks of the Florida East Coast Railroad were completely destroyed where they crossed from key to key on viaducts, and on land the rails were bodily shifted off the roadbed. The eleven-car train dispatched to Lower Matecumbe Key in an effort to rescue the veterans and other residents was stalled at Islamorada and then washed from the tracks, with only the locomotive and tender remaining on the rails.

Three populous relief camps occupied by the veterans were destroyed, and the heavy loss of life was concentrated here. The best estimates furnished by the Red Cross placed the total at 408, of whom 244 were known dead and 164 were missing. Most fatalities took place by drowning when the wall of water of the storm surge engulfed the frail structures in the camps. Some persons, however, were literally sandblasted to death and were found with no skin and no clothes except for belt and shoes. During the height of the storm, electrostatic discharges emitted from wind-driven sand appeared like millions of fireflies to give the already gruesome scene an even weirder aspect.

The cooperative observer of the Weather Bureau on Long Key, J. E. Duane, made extensive notes on the behavior of the storm as the eye passed over his fishing lodge. This excerpt provides a scientific account and a vivid description of nature at its worst:

> 9:20 p.m. — Barometer now reads 27.22 inches, wind has abated. We now hear other noises than the wind and know the center of the storm is over us. We head for the last and only cottage that I think can stand the blow due to arrive shortly. A section hand reports that a man, his wife and four children, are in an unsafe place a half mile down the track. Aid is given them and now all hands, 20 in number, are in this cottage waiting patiently for what is to come. During this lull the sky is clear to northward — stars shining brightly; and a very light breeze continued throughout lull — no flat calm. About middle of lull (which lasted a timed 55 minutes) the sea began to lift up, it seemed, and rise very fast, this from ocean side of camp. I put my flashlight out on sea and could see walls of water which seemed many feet high. I had to race fast to regain entrance to cottage, but water caught me waist deep. An idea as to rapidity with which we were inundated may be gained from the fact that the writer was about 60 feet from the doorway of the cottage. All people inside house now; water is lifting cottage from foundations. Cottage now floating.
>
> 10:10 p.m. — Barometer now 27.02 inches, wind beginning to blow from SSW.
>
> 10:15 p.m. — This was our first blast from SSW, which came full force. Seemed blast took one section of house away. House now breaking up — wind seemed stronger now than any time during storm. I glanced at barometer which read 26.98 inches, dropped it in water and was blown outside into sea — got hung up in fronds of coconut tree and hung on for dear life. Could see cottage now going to sea as parties inside were flashing a light. I was then struck by some object and knocked unconscious.

An acrimonious controversy arose over the responsibility for the presence of so many veterans on the Keys during the hurricane season. It was sparked by Ernest Hemingway, who was a member of one of the first rescue teams to reach the area. He was then living at his home on Key West, which was raked by strong winds but suffered relatively small damage. In an article entitled "Who Murdered the Vets?" appearing in *New Masses* on September 17, he loosed a blast at Washington officialdom:

> Who sent them down there?
>
> I hope he reads this — and how does he feel?
>
> He will die too, himself, perhaps even without a warning, but maybe it will be an easy death, that's the best you get, so that you do not have to hang on to something until you can't hang on, until your fingers won't hold on, and it is dark. And the wind makes a noise like a locomotive passing, with a shriek on top of that, because the wind has a scream exactly as it has in books, and then the fill goes and the high wall of water rolls you over and over and then, whatever it is, you get it and we find you, now of no importance, stinking in the mangroves.
>
> You're dead now, brother, but who left you there in the hurricane months on the Keys where a thousand men died before you in the hurricane months when they were building the road that's now washed out?
>
> Who left you there? And what's the punishment for manslaughter now?

The Great Hurricanes of 1815 and 1938 in Long Island and New England — The Great September Gale of 1815, in its northward rush from the tropics on the morning and early afternoon of September 23, bisected New England, with high winds spreading vast destruction to the east and torrential rains causing severe flooding to the west of its track. Such a destructive force was almost unprecedented. It had been 180 years since the Great Colonial Hurricane in August 1635 startled the early settlers, and only antiquarians had knowledge of that. And it would be another 123 years until a like force would strike again. This came on the afternoon and early evening of September 21, 1938, and there were few residents of the six-state area who knew more about the September 1815 Gale than its mention in a legendary poem by Longfellow.

Hurricane awareness was not part of a New Englander's conscious thinking in the 1930s, so long had it been since a major storm disaster had paid a visit. Many were to die and many to lose all their property as a result of their lack of appreciation of the massive power of a full hurricane striking directly from the sea.

The weather forecasts for Long Island and southern New England on the morning of September 21 did not contain anything to cause alarm. They called for a continuance of the cloudy, rainy conditions that had prevailed for several days. Buried in the back pages was mention of a tropical storm, earlier reported to be threatening Florida but now, after recurvature to the north, expected merely to brush the mainland and to continue a parabolic swing that would carry its wind northeast out to sea.

Actually, at sunrise on the twenty-first, the center of the whirl, now grown into a dangerous hurricane, had advanced to the Carolina Capes and lay about 75 mi (120 km) east of that checkpoint for Atlantic coastal storms. Winds at Cape Hatteras mounted to a peak of 61 mi/h (98 km/h) and the barometer dropped to 29.26 in (99.0 kPa). These figures, however, indicated a storm of only moderate proportions and did not reveal its true intensity. A later analysis showed that pressure at the center offshore was close to 27.75 in (94.0 kPa), and winds over a sizable area to the east were at 90 mi/h (145 km/h) with gusts in excess of 100 mi/h (161 km/h).

Weather forecasters at Washington issued a storm advisory at 4:00 A.M., E.D.T., placing the "hurricane" near the Carolina Capes and ordering storm warnings north to Atlantic City, New Jersey. At 10:00 A.M. the warnings were extended to Eastport, Maine, for gales up to 54 mi/h (87 km/h). At 12:30 P.M. the signals were changed to whole-gale warnings up to 74 mi/h (119 km/h) north to Sandy Hook, New Jersey. At 3:00 P.M. the advisory of the "tropical storm" warned: "Storm center will likely pass over Long Island and Connecticut late this afternoon or early tonight attended by shifting gales." This first intimation that the center might pass over a land area was received at the Providence weather bureau at 3:40 P.M. At that time the hurricane had already crossed Long Island and was well into Connecticut. Winds of 100 mi/h (161 km/h) were battering the coastal communities of Rhode Island, and a storm wave of gigantic proportions was in the process of engulfing Narragansett Bay and its shoreline.

At the early morning observation hour, the massive structure of whirling winds and waves off Cape Hatteras was thought to have a forward speed of 35 to 40 mi/h (56 to 64 km/h) and was forecast to accelerate. It was expected to be at the latitude of Long Island about 5:00 P.M. but to pass east of that exposed finger of land. The actual acceleration was about 50 percent greater than predicted, the forward speed of the hurricane being in excess of 60 mi/h (97 km/h) at its landfall on central Long Island. It smashed ashore about 2:45 P.M. in the Fire Island section of the South Shore, with the general area comprising the towns of Babylon, Bay Shore, Sayville, and Patchogue experiencing the relatively calm wind conditions prevailing for many minutes in the hurricane's eye. The minimum barometer reading reported on land was at the Bellport Coast Guard Station, 27.94 in (94.6 kPa), the lowest ever recorded on land in the 68 years of official weather records in the Northeast.

The hurricane whipped across the narrow width of Long Island, churned over the waters of the Sound, and entered Connecticut in Milford about 7 mi (11 km) west of New Haven. It then pursued a course northward over the ridges of the hills sloping down to the Connecticut River, passing west of Hartford, Springfield, and Northampton. Upon entering Vermont, west of Brattleboro, the center veered to the northwest over the spine of the Green Mountains and crossed Lake Champlain south of Burlington. By now it had lost considerable energy in its passage over land and some of its tropical characteristics, though it was still a dangerous storm, lashing the north country with high winds and drenching downpours.

The track of the 1938 hurricane when entering New England lay about 35 mi (56 km) to the west of that pursued by the September Gale of 1815, which made a landfall in the Saybrook area at the mouth of the Connecticut River. The 1815 center then moved north-by-east over the plateau of eastern Connecticut and central Massachusetts, passing between Amherst and Worcester and then close to Jaffrey and Hillsboro in New Hampshire. A wind report at Quebec next morning indicated that the center had continued curving eastward, passing to the southeast of that location.

The line of demarcation between light and heavy destruction, indicative of the path of the center of the 1815 storm, was described by the Albany-to-Boston postrider as beginning about equidistant between Springfield and Worcester: "From Brimfield to this place [Boston], there appeared to be one continued scene of devastation, in the unroofing of houses, upsetting of barns, sheds, and other buildings and in the general prostration of fences, trees, grain, and every description of vegetation." The same would apply to the 1938 countryside from about Westfield east to the western suburbs of Boston.

It was back on the coast, however, that the greatest structural damage took place and the greatest loss of human life was concentrated in 1938. Massive rises of water estimated at 15 to 20 ft (4 to 6 m) in height and topped by waves rising to crests of 30 ft (9 m) struck the open beaches of eastern Connecticut, Rhode Island, and parts of southeastern Massachusetts. All structures close to the shoreline were either washed away or severely damaged by the wall of water. In 1815, of course, there were no summer cottages on the beaches; all the population was concentrated in harbor locations where fishing and commercial boats could find anchor. The greatest damage and greatest loss of life in 1815 were reported at Stonington in extreme southeastern Connecticut. In all, about a dozen people were reported in the press to have perished, while the figure for the 1938 hurricane in this immediate shore area of Connecticut and Rhode Island numbered over 200. This reflects the growth in population and change in land use that had taken place in the intervening century and a quarter. Today there has been a tremendous increase in the summer and all-year population of these beach areas. A visit by another storm of the stature of 1815 or 1938 would be catastrophic to property and would jeopardize many lives unless warnings were heeded.

The surging storm wave accompanying a severe hurricane not only subjects the beaches to complete inundation, but it also floods into all the river estuaries and bays facing southward to the open sea. Narragansett Bay is particularly vulnerable to a storm moving inland to its west since the southerly surge of wind and tide confines the waters in its narrowing shoreline, raising the flood stage higher and higher toward the head of the bay and the diminutive Providence River. In both 1815 and 1938 the water in-

Hurricane Agnes about to strike the northeast Florida coast about noon on June 19, 1972, and enter low-pressure trough with cloud band extending far to the northeast. National Environmental Satellite Service (NOAA) photo.

vaded the center of Providence, completely halting the normal life of Rhode Island's largest city and capital. Fortunately for the historical record, someone had thought to indicate the high-water mark of 1815 on the side of a building — it read 11 ft 9.25 in (3.59 m) above mean low water. The 1938 stage topped this at 13 ft 8.5 in (4.18 m). The higher mark in 1938 was attributed to the arrival of the storm surge at the time of astronomical high tide.

Central Massachusetts and Connecticut suffered great flood damage from the tropical downpours attending the 1938 hurricane. Several days of rain had saturated the soil, setting the stage for a tremendous run-off. Hubbardston, in Worcester County, Massachusetts, received a storm total of 15.60 in (396 mm), of which 10.16 in (258 mm) fell in one day. C.C.C. Camp Buck in Portland, Middlesex County, Connecticut, measured a storm total of 17.07 in (434 mm). Swollen rivers and streams inundated residences and factories along the many watercourses. Inland, property damage from flood waters far exceeded that from the wind. The Connecticut River at Hartford rose to its second greatest flood stage of 35.4 ft (10.8 m), only 2.2 ft (0.7 m) below the all-time record established in March 1936.

The path of destruction on land in central New England in 1938 was 50 to 75 mi (81 to 121 km) wide, with lesser damage extending to 125 mi (200 km). Northeastern New Hampshire and most of Maine escaped with only light to moderate damage. The woodlands of the rest of New England, especially to the east of the storm track, suffered severely. Though damage was spotty, according to exposure and soil condition, in many places whole acres of trees were either snapped off above the ground or torn up by the roots. The total damage for Long Island and New England has been placed at $306 million, in 1938 dollars.

The exact figure for loss of life will never be known. It was placed at 600 in a later survey. The principal tragedies occurred along the Rhode Island shore and waterways, where a total of 380 died. Massachusetts placed second, with 99 fatalities, and Connecticut was third, with 85. The combined figures for deaths and property losses placed the disaster at the top of the nation's list up to that time.

Hurricane Camille's Surge on the Middle Gulf Coast in August 1969 — "By any yardstick Camille was the greatest storm of any kind ever to have affected the mainland of the United States," declared Dr. Robert H. Simpson, director of the National Hurricane Center, when assessing the data of its meteorological intensity and its physical destruction. None of those who went through the wind and water horror on the night of August 17–18, 1969, on the Mississippi and Louisiana coasts would disagree. Camille was a tightly coiled whirl of air and water whose barometric depth was exceeded only once before in American waters and whose blast force was among the highest ever measured. Only one other tropical storm in this century, the Florida Keys Hurricane on Labor Day 1935, has achieved the 5 rating on the hurricane disaster-potential scale.

Camille's landfall near midnight at a well-populated resort area in summertime was accompanied by a storm wave of surging water that rose to heights in excess of 20 ft (6.1 m), instantaneously smashing or dissolving most of man's structures facing the sea. For the duration of approximately two hours the whirl of wind and wave raged at top fury with the direction of the destructive blasts gyrating around the compass as the eye of the hurricane arrived and passed inland.

Twin destructive forces were at work. The mighty wave surge undermined foundations and swept entire buildings, along with their occupants, several blocks landward in a chaotic mass of debris in which many met a watery grave. At the same time, hurricane-force winds battered and ripped the upper stories of the surviving structures, sending death-dealing missiles flying through the air to strike down many a victim. All the while the atmosphere bellowed, the noise level of the storm attaining 120 decibels, a sound equivalent to that of a low-flying jet aircraft or a rocket engine.

The first presence of a tropical disturbance off the coast of West Africa was detected on August 5 by satellite photographs. For the next ten days the disturbed area was charted daily during its westward progress across the broad Atlantic and into the Caribbean Sea. Aircraft reconnaissance sent to check on its intensity on August 14 had the unique opportunity of witnessing by radar the formation of a storm eye and of gauging the quick intensification of its circulation. Camille was named the third tropical storm of the season. Next day, the fifteenth, it escalated to full hurricane strength, with winds estimated at 115 mi/h (185 km/h). A recurvature to the northwest took place that day, a course that crossed the extreme western tip of Cuba and laid a bead on the Gulf Coast of the United States. Once in the Gulf of Mexico the entire storm system continued to intensify and accelerated its movement toward an inevitable landfall to the north.

On the late morning of August 16 the National Hurricane

Center at Miami warned that Hurricane Camille was "small but dangerous" and was heading for the northwest Florida coast. At 5:00 P.M., the advisory reported that "small hurricane Camille becomes very intense." At 7:00 P.M., Camille was called "a very intense and dangerous storm" and was continuing to head for the Fort Walton–St. Marks area of the Florida coast. At 11:00 P.M., Camille was called "extremely dangerous" and was still threatening the northwest Florida coast. At 5:00 A.M. on the seventeenth, the "extremely dangerous" storm had shifted course westward and now endangered Mississippi and Alabama as well. Hurricane warnings were posted westward to Biloxi, Mississippi. At 7:00 A.M., they were extended as far as Grand Isle, Louisiana. At 9:00 A.M., Camille was charted as heading for the mouth of the Mississippi River. At 3:00 P.M., the center neared the southeast tip of the Mississippi delta area. By 7:00 P.M., it skirted just east of the mouth of the river and was crossing the island fringe of southeast Louisiana, about to bear down on the Mississippi–Louisiana coastline. The 11:00 P.M. advisory plotted Camille moving inland near Gulfport, Mississippi.

The center of the eye of Camille struck the Mississippi coast about 11:30 P.M. on August 17, the calm area passing over Clermont Harbor, Waveland, and western Bay St. Louis. The diameter of the eye was about 11 statute miles (18 km) and the areal diameter of the maximum force winds about 32 mi (51 km). The storm surge arrived simultaneously with the eye and reached its greatest height just to the east of the eye, a devastating 24.2 ft (7.4 m) above mean low water at Pass Christian. Barometers generally dropped below 27.00 in (91.4 kPa). An instrument later tested for accuracy located at the western end of the Bay St. Louis Bridge read 26.85 in (90.9 kPa), or only 0.5 in (1.7 kPa) higher than the generally accepted record low barometer for the United States during the Florida Keys Hurricane on Labor Day 1935.

The highest winds have been placed in the 175- to 200-mi/h (282- to 322-km/h) class. Accurate measurements were almost impossible since instruments were shattered or their supports collapsed. Aircraft reconnaissance reports offshore on the seventeenth were calculated at 175 knots or 201 mi/h (324 km/h), and little diminution took place until the eye crossed the coast. The highest surface measurement was recorded on an unattended oil rig off the Louisiana coast immediately to the east of the center; an extreme gust of 173 mi/h (274 km/h) was indicated before the paper feed of the recorder jammed. There were other estimates in excess of 150 mi/h (241 km/h) after instruments failed on account of power outages.

Maximum tides of 10 ft (3 m) or more extended from the Lake Borgne area of Louisiana near Pearlington eastward to the Alabama border near Mobile Bay, a distance of about 75 mi (120 km), and tides of 15 ft (4.6 m) or more reached from Clermont to Ocean Springs, a distance of about 37 mi (60 km). Maximum tides of 20 ft (6.1 m) were concentrated in a stretch from Bay St. Louis eastward about 22 mi (35 km) to Edgewater Park. The highest measured surge of 24.2 ft (7.4 m) came within the city limits of Pass Christian. Well to the west, at Boothville near the mouth of the Mississippi River, where the National Weather Service maintains a radar installation, the area was inundated by a tidal rise of 15 ft (4.6 m), the water standing 4.5 ft (1.4 m) deep in the weather office.

Camille's twisting wind gusts and surging waves caused a swath of destruction along the entire length of the Mississippi coast. A mass of debris was swept inland for distances varying from one to four blocks. In front of this barrier most homes and buildings were swept clean from their foundations and carried inland, and behind this line most structures remained, though severely battered by the winds. Along low-lying estuaries and bays the storm surge brought destruction far inland to buildings and boats and piled mixed debris into the marshes and woods far above the normal water line.

Highway 90, the main coastal thoroughfare, was covered with sand in some places, and in others complete sections were washed away. About one third of the Bay St. Louis Bridge and half of the Biloxi–Ocean Springs Bridge were damaged when tides lifted the spans off their supports. The Army Corps of Engineers cleared an estimated 100,000 tons of debris to make some 530 mi (853 km) of roads passable. The coast's resort industry suffered an unparalleled disaster for a major recreation area. Along Highway 90 in the Biloxi area some 60 resort properties suffered major damage, with about half of them destroyed.

At Clermont Harbor the destruction was total, and eastward

to Bay St. Louis many hundreds of beach homes were destroyed. Henderson Point, in the Pass Christian area, was completely obliterated except for an old building that was formerly a marine academy. Buildings on high ground above 20 ft (6.1 m) were able to weather the high winds and escape the storm surge, but most buildings at the 10-ft (3-m) level were swept away.

In the Gulfport area damage was severe. Three large cargo ships, the *Alamo Victory,* the *Hulda,* and the *Silver Hawk,* were badly damaged when washed ashore and grounded at the end of the harbor.

Camille's toll was high. The death list along the Gulf Coast amounted to 144 persons, with 27 listed as missing. Mississippi suffered 135 fatalities and Louisiana 9. The total damage was estimated at $1420 million, about three quarters of the sum being in Mississippi. Alabama reported no deaths and relatively light damage.

October

October gave a party;
  The leaves by hundreds came:
The ashes, oaks, and maples,
  And those of every name.
    — George Cooper, "October's Party"

O suns and skies and clouds of June,
  And flowers of June together,
Ye cannot rival for one hour
  October's bright blue weather.
    — Helen Hunt Jackson, "October's
    Bright Blue Weather"

I bow me to the threatening gale:
  I know when that is overpast,
Among the peaceful harvest days
  An Indian Summer comes at last.
    — Adeline Dutton Train Whitney,
    "Equinoctial"

An Indian Summer comes at last!" October is the truly transitional autumnal month, starting out with summery green and ending with wintry gray. But often there is a pause in the inexorable advance of the season, when a throwback to summer comes and a period of pleasant weather prevails, with mild temperatures, light winds, and a hazy atmosphere. The concept of Indian summer has intrigued a variety of people from romantic dreamers to pedantic professors. Many a poet has tried his hand at creating verse about its glories, and philologists have probed the origin and usage of the term. In the opening years of the early 1900s, Albert Matthews, a Boston lexicographer, perused practically the entire mass of early American literature in search of the first use of the term. He produced a lengthy dissertation, which was published in *The Monthly Weather Review,* but could not determine the region of its origin or exact meaning beyond dispute.

Meteorologists are generally agreed that the "second summer" or "fifth season" is brought on by the stagnation of an anticyclone that continues over an area for several days in a row. Descending warm air aloft, known as the process of subsidence, creates a temperature inversion, with warm air overlying colder air at the surface. This puts a lid on the atmosphere and prevents the normal upward venting of the impurities that man and nature release into the air. Warming hazy sunshine characterizes Indian summer days, and one enjoys a relaxed feeling in the softness of the weather. Some claim that Indian summer cannot come until after the first frost of autumn or a period of freezing weather known as "Squaw Winter." It is a floating period on nature's calendar, coming anytime in October or November, and there may be a recurrence of the pleasant spells two or three times during the same autumn.

Many sections of the country experience conditions resembling Indian summer, though the Middle Atlantic States and the Ohio Valley are best known for the frequency of its occurrence. This is where the term first appeared in written literature about the time of the Revolution, and pioneers carried its reputation to the far corners of the country. What is now known as the West Virginia High is the responsible meteorological agent. Since this region is also the location of much manufacturing activity, Indian summer now carries the threat of air pollution episodes that may be injurious to one's health. So it is regarded now as a mixed blessing.

The annual contraction southward of the Pacific High becomes evident during the month of October as the central axis retreats to 33°N, the latitude of San Diego, and the weakened northeast arm is no longer the impenetrable blocking factor for Pacific storms. Meanwhile, the Aleutian Low has assumed a firm position astride the southwest Alaskan mainland and inner Aleutian Islands; its central mean pressure decreases to 29.60 in (100.3 kPa). All is now ready for the winter work of manufacturing storms. The thermal low over the Southwest of the United States has weakened in strength and withdrawn its summer thrust from the Great Valley of California and interior Oregon.

In the Atlantic Ocean, the Azores–Bermuda High remains in the latitude of Charleston, South Carolina; but its center has shifted to the east over the ocean, leaving a small subcenter over Virginia and the Carolinas. The Iceland–Greenland Low exhibits a split, with a primary center in Davis Strait and a secondary center off south Iceland, forewarning of more frequent storm activity across the North Atlantic Ocean.

The jet stream, remaining at the northerly latitude of September, enters the continent over southern British Columbia before traversing the southern Prairies and dipping slightly southeast across the northern Great Lakes, and then pursuing a course east to the Atlantic Ocean south of Newfoundland.

During October the main belt of westerly flow begins to move south. A corresponding displacement of the principal storm tracks becomes evident in Canada, where the primary cyclonic path crosses James Bay rather than central Hudson Bay. It is joined there by a new seasonal track leading from the central Great Plains northeast over Lake Michigan and into Ontario. The tropical storm zone lies farther off the Atlantic coastline than in September, and the track north from the Gulf of Mexico crosses north Florida and Georgia into the Atlantic. Some mighty hurricanes have followed this route and caused extensive damage north to New England.

The main path of Pacific anticyclones enters the continent over Oregon and trends east to South Dakota, where it merges with a path of polar highs from western Canada. Two routes continue east from the northern Plains: One leads over the Lower Lakes, then turns northeast over New England; the other swings south as far as north Arkansas and Tennessee before recurving northeast to pass over West Virginia and to the ocean south of Long Island. High-pressure areas show a tendency to move slowly in October and often stall in the Ohio Valley and the central Appalachians for two to three days. This is the situation that produces the delightful autumn days of Indian summer.

Except for some small isolated islands of heavier precipitation, all of the eastern half of the country receives from 2 to 4 in (51 to 102 mm) of rain in October, often the driest month of the year. But in eastern and southern Florida, where tropical storms often prevail, an 8-in (203-mm) rainfall is normal from Palm Beach southward. The Rocky Mountains are drier than in September, but the Intermountain region shows a slight increase in precipitation. The rainy season has definitely begun in the Pacific states with the 8.0-in rainfall line appearing in the Coastal Range and the Cascades of Oregon and Washington, and the 4.0-in (102-mm) area extending south, well into California.

Gone from the temperature distribution map is the 80°F (27°C) mean isotherm, and the plus-70°F (21°C) area is restricted to Florida, the immediate Gulf Coast, the southern half of Texas, and the lowlands of the desert Southwest. The bulk of the South experiences temperatures between 60 and 70°F (16 and 21°C), while the North lies mainly in the 50–60°F (10–16°C) range. Northern New England, northeastern New York, and the Canadian border region westward have a mean temperature below 50°F (10°C) in October. There is a general cooling throughout the West, especially in the Valley of California, where temperatures are in the 60–70°F (16–21°C) zone. The October mean at mountain exposures falls below 50°F (10°C).

Temperature extremes for October: 116°F (47°C) at Sentinel, Arizona, October 5, 1917; minimum −33°F (−36°C) at Soda Butte, Wyoming, October 29, 1917.

The more important farm concerns during October are the

seeding of winter grains, the husking or picking of corn, and the harvesting of cotton and late fruit. The threshing of spring grains is usually completed during the month, and also the seeding of winter wheat, rye, and oats in most sections. The bulk of the cotton crop is usually picked by the last of the month. Livestock remain on the range in most western districts but are frequently moved to lower levels, while in some localities cattle and sheep are driven to ranches.

October

1 **1752** Second severe hurricane in two weeks struck Carolinas; destroyed Onslow County courthouse with all records; Beacon Island disappeared; some blamed the disaster on the change from Julian to Gregorian calendar in early September.

2 **1882** Severe early-season windstorm over Northern California and Oregon; great crop damage in Sacramento Valley; thousands of trees blown down.

3 **1841** The October Gale, Nantucket's worst; caught Cape Cod fishing fleet at sea, 57 men lost from Truro alone; 40 ships ashore on Cape Cod; heavy snow inland, 18 in (46 cm) near Middletown, Conn.

4 **1777** Battle of Germantown opened with morning fog that soon mixed with the smoke of battle; poor visibility caused confusion; Americans fired on each other, contributing to loss of battle.

1869 Saxby's Gale and Great New England Rainstorm and Flood; storm predicted twelve months ahead by British officer; great tide and wind in Maine and New Brunswick; very heavy rain and high floods in all New England; 12.35 in (314 mm) at Canton, Conn.

5 **1638** "A mighty tempest, and withal the highest tide, which has been since our coming into this country" (Gov. John Winthrop). Second severe hurricane in three years; blew down many trees in mile-long tracks.

1786 Famous Pumpkin Flood on Susquehanna and Delaware rivers; high stage of 22 ft (6.7 m) at Harrisburg; wet season culminated in heavy showers.

6 **1836** Second heavy 1836 snowfall: Auburn, N.Y., 26 in (66 cm); Wilkes-Barre, Pa., 11 in (28 cm); all mountains in Northeast covered.

7 **1970** Puerto Rico suffered "most widespread natural disaster in recent years" when floods struck all parts; 38.42 in (976 mm) rain fell in six days; $62 million damage.

8 1871 Great Chicago Fire; severe drought had prepared tinder-dry scene; southwest winds spread fire across city; 250 persons perished; damage $196 million; on same night a forest fire roared into Peshtigo, Wis., killing an estimated 1000 residents.

9 1703 General early-season snowstorm from Philadelphia to Boston. "The snow is now three to four inches deep . . . a sad face of winter" (Judge Samuel Sewall's diary).

1903 New York City's heaviest rainstorm, with 11.17 in (284 mm) measured in Central Park in 24 hours; severe flood in New Jersey's Passaic Valley, where over 15 in (381 mm) fell.

10 1804 Famous Snow Hurricane; unusual storm caused northerly gales from Maine to New Jersey; heavy snow in New England with 36 in (91 cm) at crest of Green Mountains; 12 in (30 cm) in Berkshires at Goshen, Conn.

1973 Deluge of 15.68 in (392 mm) at Enid, Okla., flooded city.

11 1925 Widespread early snow on the October 10–11 weekend; blocked roads, canceled football games; 24 in (61 cm) in Vermont and New Hampshire.

1954 Chicago deluge of 6.72 in (171 mm) in 48 hours; Chicago River in flood.

12 1836 Third heavy 1836 early snow, with 18 in (46 cm) at Bridgewater, N.Y., and 12 in (30 cm) at Madison, N.Y.; mountains in Northeast whitened for third time.

1962 Columbus Day "Big Blow" in Oregon and Washington; winds in excess of 100 mi/h (160 km/h); 28.42 in (92.2 kPa) barometer; 10 million board feet blown down in forests; extensive structural damage; 48 deaths; $210 million loss.

13 1846 Great Hurricane of 1846; track: Cuba–Key West–Florida peninsula and through Georgia, South Carolina, North Carolina, Virginia, and Pennsylvania; major damage in all areas; similar to Hazel in 1954.

14 1965 Ft. Lauderdale, Fla., had deluge of 24 in (610 mm) in 24 hours for local intensity record.

15 1954 Hurricane Hazel roared inland at Myrtle Beach, S.C.; wind at 106 mi/h (170 km/h); barometer at 28.47 in (96.4 kPa); vast damage on coast and inland through North Carolina, Virginia, Maryland, and Pennsylvania; $250 million loss.

16 1880 Early-season blizzard struck Dakotas and Minnesota; railroads blocked; drifts remained throughout severe following winter; gales did extensive shipping damage on Great Lakes.

17 1910 Loop Hurricane off southwest Florida hit Ft. Myers area; Sand Key had barometer of 28.40 in (96.2 kPa) with winds over 100 mi/h (160 km/h).

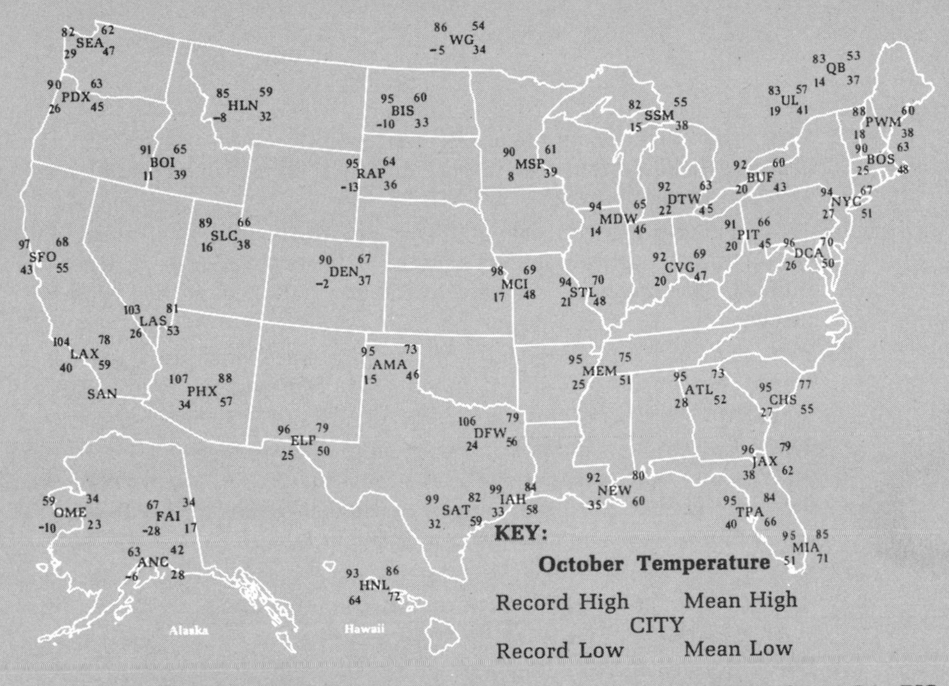

Station Designators: AMA Amarillo TX; **ANC** Anchorage AK; **ATL** Atlanta GA; **BIS** Bismarck ND; **BOI** Boise ID; **BOS** Boston MA; **BUF** Buffalo NY; **CHS** Charleston SC; **CVG** Cincinnati OH; **DCA** Washington DC; **DEN** Denver CO; **DFW** Dallas-Fort Worth TX; **DTW** Detroit MI; **ELP** El Paso TX; **FAI** Fairbanks AK; **HLN** Helena MT; **HNL** Honolulu HI; **IAH** Houston TX; **JAX** Jacksonville FL; **LAS** Las Vegas NV; **LAX** Los Angeles CA; **MCI** Kansas City MO; **MDW** Chicago IL; **MEM** Memphis TN; **MIA** Miami FL; **MSP** Minneapolis St. Paul MN; **NEW** New Orleans LA; **NYC** New York NY; **OME** Nome AK; **PDX** Portland OR; **PHX** Phoenix AZ; **PIT** Pittsburgh PA; **PWN** Portland ME; **QB** Quebec QUE; **RAP** Rapid City SD; **SAN** San Diego CA; **SAT** San Antonio TX; **SEA** Seattle WA; **SFO** San Francisco CA; **SLC** Salt Lake City UT; **SSM** Sault Ste. Marie MI; **STL** St. Louis MO; **TPA** Tampa FL; **UL** Montreal QUE; **WG** Winnipeg MAN; **YC** Calgary ALB.

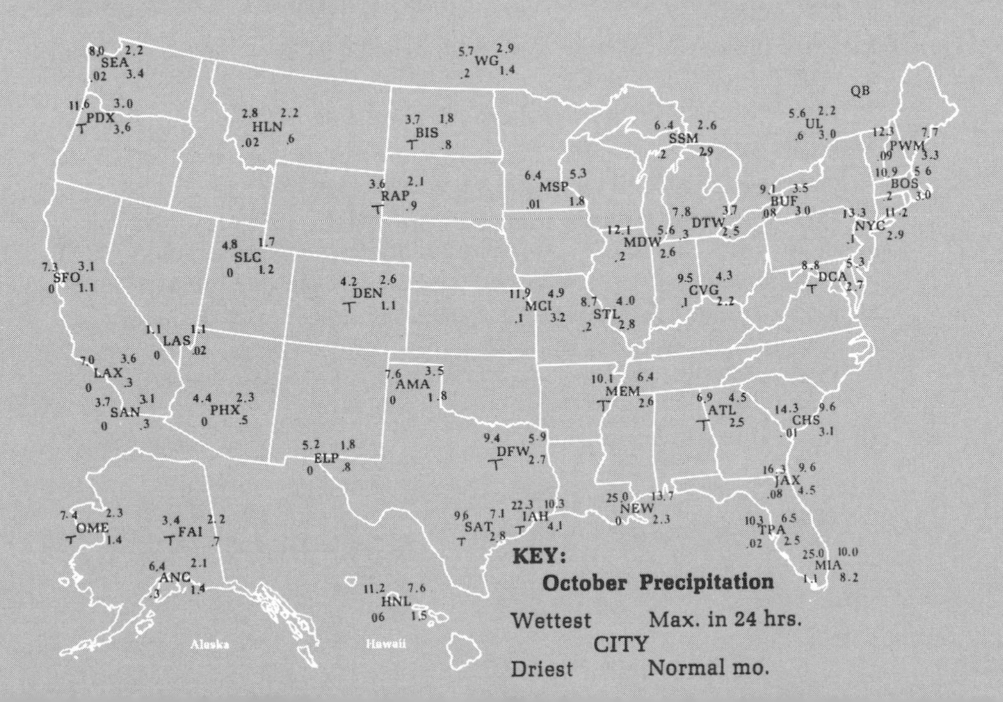

18 1930 Big early-season, lake-effect snowburst on lee shores of Lake Erie and Lake Ontario; 48 in (122 cm) fell just south of Buffalo and also at Gouverneur, N.Y.

19 1844 Famous Lower Lakes Storm of 1844; southwest hurricane winds blew for five hours, driving lake waters into downtown Buffalo; Toronto had barometer of 28.86 in (97.7 kPa); 200 reported drowned.

20 1770 Eastern New England experienced "an exceedingly great NE storm"; extensive coastal damage from Massachusetts to Maine; highest tide since 1723.

21 1492 Christopher Columbus made landfall on San Salvador Island under clear skies; fortunately, he experienced no hurricanes on first voyage and returned safely to Spain in March 1493, though *Santa Maria* was wrecked on reef off Cuba.

22 1965 Heat wave of record duration hit Southern California; Los Angeles had 100°F (38°C) or more from October 20 to 29; San Diego at 104°F (40°C) on October 22.

23 1761 Southeast New England Hurricane, "most violent in thirty years"; thousands of trees uprooted in Massachusetts and Rhode Island, blocking roads.

24 1785 Four-day rains put Merrimack River in New Hampshire and Massachusetts to greatest height then known; extensive damage to bridges and mills.

1878 Hurricane caused widespread losses in North Carolina, Virginia, Maryland, New Jersey, and Pennsylvania; "Philadelphia's worst"; barometer 28.82 in (97.6 kPa) at Annapolis.

1947 Bar Harbor holocaust in Maine when wind-driven forest fires consumed homes and medical research institute; 17 died; damage $30 million.

25 1919 Bismarck, N.D., recorded its earliest below-zero reading with −10°F (−23°C).

26 1859 New York City's earliest substantial snowfall of 4 in (10 cm).

27 1764 "Very remarkable storm of snow with high wind"; 22 in (56 cm) reported at Rutland in central Massachusetts.

28 1971 Severe early-season blizzard over Plateau and Rockies; heavy snows with 27 in (69 cm) at Lander, Wyo.; record cold, −15°F (−26°C) at Big Piney, Wyo.; Interstate Highways, railroads blocked.

29 1942 Tornado at Berryville, Ark.; 29 killed; 137 buildings destroyed; damage $500,000.

1917 Soda Butte, Wyo., registered −33°F (−36°C), lowest for U.S. in October; Denver dropped to 0°F (−18°C).

30 1947 Donora Smog Disaster caused 20 deaths in Monongahela

Valley of Pennsylvania when five-day inversion trapped impurities in lower atmosphere.

31 1965 Ft. Lauderdale received 14 in (356 mm) more rain, giving a month's total of 42.43 in (1078 mm).

Air Pollution

Pollution of our atmosphere is not a new problem. It arises both from natural sources and from the activities of man. Volcanoes, forest fires, decaying vegetation, bacterial emissions, and dust storms have always contributed impurities to the atmosphere, but the worldwide dispersion of their products results in a low average concentration in a particular locality. This is not the case with manmade pollutants, which have a tendency to concentrate near our urban complexes and thus affect a large number of humans.

Urban air pollution has been mentioned since Roman times. In Western Europe, measures were taken in the late Middle Ages to control smoke emissions, and the industrial revolution in the nineteenth century greatly exacerbated the situation by adding other impurities to the atmosphere. This type of pollution has been called "classical smog," the apt designation being an abbreviated combination of the words *smoke* and *fog*. It was first employed in 1905 to describe the constituents of some of London's famous "pea-soup" fogs. American cities, too, experienced troublesome carbon combustion problems in the late nineteenth and early twentieth centuries when coal was the principal energy source of fuel and power. It was considered a nuisance but not a major problem.

It was not until after World War II that the effects of air pollution on many aspects of modern life became clearly understood. In the Los Angeles basin of California the recognition of "photochemical smog" and the identification of automobile exhaust as the principal contaminating agent came in the early 1950s. Since then, with the great increase in vehicular traffic and industrialization, smog problems have plagued most urban concentrations of population and have required an attack organized on a national scale.

A large number of substances employed in industrial and commercial operations have been found to be toxic, and the venting and discharge of such materials into the atmosphere is either prohibited or controlled. Of the large number of potentially dangerous substances, only a few are present in the atmosphere in sufficient quantities to be of immediate concern. Particulate matter, carbon monoxide, sulfur oxides, oxides of nitrogen, unburned hydrocarbons, photochemical oxides, asbestos, berylium, and mercury were considered to be sufficiently hazardous to warrant the establishment of either air-quality standards or emissions regulations by the federal government.

Currently, the most serious problems are presented by the reaction of certain chemical agents in the atmosphere. Two types have drawn national attention: One is characterized by a high concentration of sulfur oxides in the air as a result of burning sulfur-bearing fuels in industrial processes or in home heating. Conditions are aggravated by a high moisture content of the air along with a concentration of suspended particulate matter in the atmosphere. Designated "sulfurous pollution," it appears over many Eastern and Midwestern cities, where it has been found to be injurious to human respiratory organs, damaging to plant and wildlife, and prone to attack the masonry of buildings. The sulfur-bearing particles, when carried downwind, serve as the nuclei for the formation of raindrops and fall to the ground as "acid rain" some hundreds of miles from the emitting source. The problem has received much attention of late because it has been demonstrated that streams and lakes in formerly pristine mountain country have become polluted from afar and local fishlife has been extinguished.

A second type of chemical reaction involves photochemical processes created by sunlight. Incomplete combustion of automotive fuel produces nitrogen oxide, which reacts with hydrocarbons under sunlight, ultimately producing ozone and associated irritants. The main effect on humans is the production of eye irritation and respiratory afflictions along with other long-range

Opposite: A smog's eye view: Where's New York? October 27, 1963. Courtesy of Environmental Protection Agency.

health side effects. Such pollution also attacks crops and vegetation, with serious results. Overall, it develops an unattractive yellow-brown tinge to the atmosphere that restricts visibility and creates an unpleasant environment, both physically and esthetically.

Weather Factor

The structure of the atmosphere in its lower levels is a major factor in creating and intensifying air pollution situations. The atmosphere serves as a mighty disposal system that can take care of enormous amounts of pollutants and cleanse itself. But when a lid is put on its lowest strata, the impurities are trapped near the surface of the earth and cannot vent upward. This is known as an inversion, whereby a layer of warm air overlies a layer of cool air hugging the surface. Normal upward convective currents cannot penetrate the inversion, so smoke, automobile exhaust, and other pollutants spread out laterally over the countryside, carrying pollution downwind.

Inversions occur most frequently during high-pressure periods of good weather, with clear skies, light winds, and stagnant circulation. Inversions form during the overnight cooling, and often dissipate by midmorning when the heating of the sun penetrates to earth, warms the cool layer, and convective currents wipe out the inversion. Sometimes the inversion layer is too thick or too turbid to permit the sun's rays to penetrate; then the inversion may become semipermanent until the arrival of a weather front and a new air mass.

The West Coast is noted for its almost daily inversion formations in the warmer months of the year. The waters of the Pacific Ocean cool the lower layer of the atmosphere, and this moves onshore with the sea breeze. Warmer air over the land is displaced upward, creating the inversion. Pollutants are vented into the lower layer as it drifts farther inland. Local topography plays an important role in intensifying a smog situation; in the Los Angeles area high mountains to the north and east form a basin and prevent the lateral spread of the polluted air, so it is channeled farther and farther inland during the day, sometimes as far as 60 mi (97 km). But in the evening a land breeze carries the smog seaward. This cycle, with more pollutants being added daily, may continue for several days until broken up by the arrival of a weather system to change the entire regional circulation pattern and cleanse the air.

Air-Pollution Incidents

Several times in the past few decades atmospheric conditions in industrial and urban areas have become so adverse to human health as to cause an above-average number of deaths. The narrow valley of the Meuse River west of Liège, Belgium, experienced a deadly industrial smog during the opening days of December 1930 when at least sixty-three persons died and hundreds were stricken with respiratory ailments. The city of London was subjected to severe air-pollution conditions in 1952, 1953, 1956, and again in 1962. The first incident was by far the most serious. In two weeks more than 4000 persons were hastened to their graves by a cold atmospheric fog that combined with effluents to cause the catastrophe.

In the United States the first incident to receive widespread notice occurred in the valley of the Monongahela River at Donora, Pennsylvania, about 20 mi (32 km) south-southeast of Pittsburgh. During the last week of October 1948 a thick blanket of polluted air hung over the valley; the near-stagnation became so aggravated that in a fourteen-hour period seventeen deaths occurred when only one might be expected from normal mortality rates. An anticyclonic regime dominated the area from October 26 to 31, until a cold front came along on Halloween to clear the air. In all, nineteen died, all but two at the height of the stagnant conditions on the thirtieth. Most were elderly people who suffered from cardiac or respiratory disease. About 43 percent of the population of Donora experienced some effects from the smog; they complained of such upper respiratory symptoms as nasal discharge, constriction of the throat, or sore throat.

Similar atmospheric and pollution situations had developed on at least two previous occasions at Donora: from October 14 to 18, 1923, and October 7 to 18, 1938. This is the season of the famed Indian summer in the region, a period of the year that formerly delighted the residents for its gentle weather, but now carries the threat of discomfort and even death.

Smoggy day in Boston. Boston Globe photo.

New York City has endured an unusually polluted atmosphere rather frequently of late, episodes taking place in 1952, 1953, 1956, 1962, 1963, 1966, and 1970. The first occurred from November 12 to 23, 1952, when more than 175 persons died of cardiac and respiratory distress. The Thanksgiving Day incident of 1966 resulted from the stagnation of a high-pressure area from November 20 to 25. Pollution reached a peak on the twenty-fourth, Thanksgiving Day, when the carbon dioxide level rose from 8 parts per million on the twenty-second to 35 ppm. Oxides of nitrogen, sulfur, and hydrocarbons also showed maximums. An increase of twenty-four deaths per day was reported during the period.

The city of Los Angeles, where some people think smog was "invented," experiences the most frequent air-pollution conditions as a result of its geographic situation, stable weather regime, and excessive number of automobiles. The first intimation of the harmful effect of air pollution came in 1942, when damage to plant life was first observed. The first actual detection of a smog condition occurred on September 8, 1943. No doubt the great increase in industrial activity and transport movement attendant on World War II filled the air with contaminants beyond usual limits. In 1947 the Los Angeles County Air Pollution Commission was established, and its subsequent program has been outstanding in the preventative field, as well as in professional research into the nature and behavior of air pollution.

The United States Government moved in 1963 with a Clean Air Act to protect its citizens against air pollution that "endangers the health or welfare of any persons." This was strengthened with the enactment of amendments in 1970. In the Clean Air Act of 1977, the United States Congress endorsed the concept of a nationwide air pollution reporting system based on a nationally standardized scale.

SP

November

If there's ice in November that will bear a duck,
There'll be nothing after but sludge and muck.

Ice in November
Brings mud in December.

If ducks do slide at Hallontide,
At Christmas they will swim;
If ducks do swim at Hallontide,
At Christmas they will slide.
(Hallontide = Halloween-tide, November 1.)

If All Saints' Day [November 1] will bring out winter, St. Martin's Day [November 11] will bring out Indian summer.
— Richard Inwards, Weather Lore (1898)

Opposite: The Pacific fleet enters San Francisco Bay through fog-shrouded Golden Gate and passes under the Bay Bridge to Oakland in clear weather. U.S. Navy photo.

"No shade, no shine, no butterflies, no bees, No fruits, no flowers, no leaves, no birds,—No-vember!" Thomas Hood wrote of nature's negative aspects in his poem "No." But the meteorological realm is in anything but a negative mood during the eleventh month when some of the mightiest cyclonic storms have struck. The Anglo-Saxons called it the "winde-monath" in their isle, and it has lived up to its reputation on this continent. In the days of sailing ships on the Great Lakes, navigation was usually suspended by the end of October after the tragic storm of November 11, 1835, which "swept the lakes clear of sail." In the present century there have been two notorious lake storms on November 8–9, 1913, and November 11, 1940, which brought tragedy to many sailors. Though the threat of a hurricane is greatly diminished, the Atlantic Coast can experience severe northeasters at this season as witnessed by the Portland Storm on November 26–27, 1898, and Great Appalachian snow-and-windstorm on November 25, 1950. Along the Northwest Coast old residents remember the wind, ice, and snowstorm on November 19, 1921, which filled the highway and railroad passes with 60 in (152 cm) of snow, blocking all traffic through the Columbia River gorge for several days. Only the Gulf Coast seems immune from nature's mighty November outbursts.

The Pacific High continues the retreat to the south, reaching a position off northern Mexico, and reduces its strength to a winter normal of 30.12 in (102.0 kPa). Storms can now move across the north Pacific Ocean without hindrance from the Pacific High and strike the coast of California and northward. A subcenter of high pressure develops over adjacent areas of Idaho, Nevada, and Utah; this will become an increasingly important feature of early winter weather maps.

The Aleutian Low moves into its winter home in the northwest corner of the Gulf of Alaska and deepens to a mean pressure of 29.53 in (100.0 kPa). Periodically it will dispatch a series of low-pressure impulses to the mainland, whose storm centers will drop heavy rains on coastal sections and deep snows in the mountains.

In the Atlantic Ocean, the Azores–Bermuda High holds in

the same vicinity as in October and retains its strength, although the autumnal subcenter over the Carolina–Virginia mainland disappears. This opens the way for coastal storms to sweep northward along the Atlantic seaboard. The Iceland and Greenland low centers are now joined in a broad zone of low pressure; the mean pressure is down to 29.60 in (100.3 kPa), indicative of increased storm activity in the North Atlantic as the winter solstice approaches.

After crossing the mid-Pacific Ocean close to the 40°N parallel, the jet stream swings north to enter the continent over southern British Columbia, and then continues east in an undulating flow, dipping into the Midwest as far south as the Ohio Valley before swinging northeast over New England.

In November the storm tracks take on paths that are characteristic of the cold season. In the eastern Pacific Ocean three winterlike paths reappear on weather maps and lead to maximum storm frequency off the coast of British Columbia, with frontal passages inland over the American Northwest. Another new aspect of the maps in November is the convergence of the tracks of Alberta and Colorado lows toward the upper Great Lakes, where a second area of maximum storm frequency is found. This will remain throughout the winter months and account for the severity of the season in that region.

The majority of anticyclonic tracks also take on typical wintertime locations. In the interior of North America, the trajectory from north to south of polar highs from Canada assumes primary importance as a weather-controlling factor for the first time since March. Most Canadian polar highs move southeast on a well-defined track carrying west and south of the Great Lakes, but a few Canadian highs move north of the Great Lakes, through Ontario and southern Quebec, before leaving the continent over Maine and the Atlantic Provinces. High pressure is pronounced in the Intermountain region of the West; offshoots occasionally move southeast through the Southern States to the Appalachians.

Below-freezing mean temperatures appear in November along the northern border from Montana eastward to Wisconsin. Higher elevations in the Northeast will also have means below freezing, as does most of the higher ground in the Rocky Mountains. The

50°F (10°C) isotherm runs from the lower Chesapeake Bay southwest across the northern parts of the Gulf States to take in all but the panhandle of Texas. The 60°F (16°C) area includes the peninsula of Florida, the delta of Louisiana, and southeast Texas. Only the southern coasts of Florida have means above 70°F (21°C). It is much cooler everywhere in the West. The desert valleys of the Southwest average above 60°F (16°C), but elsewhere it is generally below 60°F over most of California and below 50°F (10°C) in coastal Oregon and Washington.

Temperature extremes for November: 105°F (41°C) at Croftonville, California, on November 12, 1906; minimum −53°F (−47°C) at Lincoln, Montana, on November 16, 1959.

The winter precipitation pattern in the South appears with the lower Mississippi Valley exceeding 4.0 in (203 mm) north into Kentucky. Coastal and southern New England and the Middle Atlantic Coast also have a rainy month with more than 4.0 in. The Great Plains are drier than in October with the one-inch (25-mm) line running north and south from western Minnesota to central Texas. The rainy season extends its hold southward along the Pacific Coast to Southern California. The 8.0-in (203-mm) area reaches into Northern California, and as many as 16 in (406 mm) are normal in the Olympic Mountains of Washington.

The most important farm activities in November are gathering corn, picking and ginning cotton, and marketing potatoes and fall truck crops. In southern sections, the seeding of winter grains and the planting of fall and winter gardens continue. The digging of sugar beets is still in progress in the Rocky Mountain and Intermountain regions, and by the close of the month the bulk of the crop has been delivered to factories. In most of the western-range sections, livestock are moved from higher elevations to the lower plains or feed lots. In the north Pacific area, pastures usually are greatly improved because of the increased autumn rainfall.

November

1 1861 Cape Hatteras hurricane battered Union fleet attacking Carolina ports, later raised high tides and winds in New York and New England.

2 1743 Ben Franklin's "Eclipse Hurricane" unlocked key to coastal storm movement; B.F. at Philadelphia prevented from viewing lunar eclipse in northeast rainstorm, but his brother at Boston saw it, though rain began an hour later there.

3 1890 Los Angeles hit 96°F (36°C), long a November record until maximum of 100°F (38°C) on November 1, 1966.

4 1927 Great Vermont Flood; two-day rain of up to 9.0 in (229 mm) put rivers in west New England over banks; Winooski Valley devastated; 84 dead in Vermont, 88 in all New England; $45 million loss.

5 1894 Famous "Election Day Snowstorm" began in Connecticut; wet 10–12-in (25–30-cm) snow caused great damage to wires and trees in high winds; 60 mi/h (97 km/h) peak on Block Island.

1977 Heavy rains caused dam break, flood at Taccoa, Ga.; 38 dead.

6 1951 Early-season snowfall in Missouri; 13.3 in (34 cm) was greatest so early in season at St. Louis; Washington County measured 20 in (50 cm).

7 1940 "Galloping Gertie," the Narrows Bridge at Tacoma, Wash., collapsed when near-gale winds caused whole structure to vibrate excessively.

8 1870 First storm warning by U.S. Army Signal Service issued for Great Lakes region by Prof. Increase Latham at Chicago.

1966 Downtown San Francisco hit 86°F (30°C) for highest ever in November.

9 1913 "Freshwater Fury" swept Great Lakes in famous storm; eight large ore-carriers went down on Lake Huron with 200 lives lost; Erie barometer dropped to 28.61 in (96.9 kPa); Buffalo wind at 62 mi/h (100 km/h); heavy snow: 22.2 in (56 cm) at Cleveland, 36 in (91 cm) at Pickens, W. Va.

1926 Tornado in Charles County in southern Maryland killed 17 people.

10 1915 Tornado at Great Bend, Kan., killed 11 people, did $1 million damage.

1975 Another "Freshwater Fury" on Lake Superior sank ore-carrier *Edmund Pendleton* with loss of crew of 29.

11 1940 Armistice Day Storm in Upper Midwest and Great Lakes; blizzard in Minnesota and Wisconsin; 49 dead in Minnesota alone; whole gales on Lake Michigan caused wrecks and loss of 59 sailors; 17-in (43-cm) snowfall in Iowa; barometer at 28.66 in (97.1 kPa) at Duluth.

1955 Early arctic outbreak into Washington and Oregon damaged shrubs and fruit trees; many new November low-temperature records set: 0°F (−18°C) in western, −19°F (−28°C) in eastern Washington.

12 1906 Craftonville, Calif., hit 105°F (41°C), hottest ever for U.S. in November.
1974 Great Alaska Storm in Bering Sea caused worst coastal flooding remembered at Nome, with tide 13.2 ft (4 m); damage estimated at $12–$15 million; no lives lost.

13 1833 Great Meteor Display when Leonids put on spectacular show from midnight to dawn under generally clear skies in most sections; repeated in western U.S. in 1966.
1933 First great "Dust Bowl Storm" spread pall from Great Plains to New York State.

13 1946 Initial cloud-seeding experiment with dry ice by Schaefer and Langmuir over Massachusetts Berkshires converted 4-mi cloud into snow flurries.

14 1972 Coastal storm gave most of New England a lashing; heavy wet snow up to 18 in (46 cm) in Vermont and New Hampshire downed utility and communication lines; heavy rain with flooding in south; high surf and erosion along coast.

15 1900 Record lake-effect snowstorm at Watertown, N.Y., with 45 in (114 cm) in 24 hours.

16 1959 Severest November cold wave in history; Lincoln, Mont., dropped to −53°F (−47°C); Helena had a record snowfall of 21.5 in (55 cm).

17 1869 Southwest winds of hurricane force swept Berkshires and Green Mountains, doing extensive forest and structural damage.
1927 Tornado crossed Alexandria, Va., and southeast Washington, D.C.; struck Naval Air Station; path 17 mi (27 km) long; gusts of 93 mi/h (150 km/h); waterspout occurred 90 minutes later.

18 1873 Severe storm raged from Georgia to Nova Scotia; Portland barometer at 28.49 in (96.5 kPa); great losses to fishing fleets.

19 1921 Severe storm in Northwest; The Dalles, Ore., had 54 in (137 cm) of snow and sleet, blocking Columbia River highway; railroads stopped for days.

20 1869 Second great windstorm in three days in Vermont and New York blew railroad train from tracks.
1900 Tornado outbreak in Arkansas, Mississippi, and Tennessee killed 73 people and did extensive material damage.

21 1798 "The Long Storm" dropped 12 in (30 cm) of snow on New York City and New Haven; up to 36 in (91 cm) fell in New Hampshire and Maine in four-day storm; long, severe winter followed.
1967 Excessive rains in Southern California with 14.0 in (356 mm) in mountains and 7.96 in (202 mm) in downtown Los Angeles;

severe local flooding with damaging mudslides; "worst since 1934."

22 1641 Boston: "A great tempest of wind and rain from the S.E. all the night, as fierce as a hurricane... and thereupon followed the highest tide which we have seen since our arrival here" (*Journal of John Winthrop*).

23 1943 Record snowfall in north New Hampshire: Randolph 56 in (142 cm); Berlin 55 in (140 cm); many others exceeded 40 in (102 cm).

1960 Tiros II launched; had life of 10 months and took 25,574 photos.

24 1812 Southwest hurricane-force winds at Philadelphia and New York City unroofed buildings and sank ships.

1863 "Battle above the Clouds" on Lookout Mountain near Chattanooga; pre-frontal clouds obscured upper battlefield, aiding Union victory.

25 1926 Tornado swept through Belleville and Portland, Arkansas, killing 53 and doing $630,000 damage.

1950 Great Appalachian Storm in Northeast; heavy snows on western slopes: 57 in (145 cm) at Pickens, W. Va.; excessive rain on eastern slopes: 7.78 in (198 mm) at Slide Mountain, N.Y.; 108 mi/h (174 km/h) at Newark, N.J.; record November cold in Midwest and Upper South.

26 1888 Late-season hurricane passed close to Nantucket and Cape Cod, later crossed Nova Scotia; heavy shipping losses and beach erosion.

1896 Thanksgiving Day blizzard in North Dakota; "wind velocity or snowfall never equaled before."

27 1898 "Portland Storm" named after coastal passenger ship that foundered off Cape Cod with loss of all aboard, about 200 people; many small vessels sank with about 200 more casualties; whole gales on coast and heavy snows inland, 27 in (69 cm) at New London, Conn.

28 1960 Severe storm on Lake Superior raised waves 20–40 ft (6–12 m) high: 73 mi/h (117 km/h) gusts; north shore flooded; erosion and damage.

29 1921 Central New England had four-day ice storm; worst in modern records; Worcester area hard hit; trees and wires downed; damage in milions.

30 1875 Severe early cold wave set November marks in Northeast: New York City, 5°F (−15°C), Boston, −2°F (−19°C), Eastport −13°F (−25°C).

1896 Severe November cold wave in Upper Midwest: −45°F (−43°C) at Pokegama Dam, Minn.

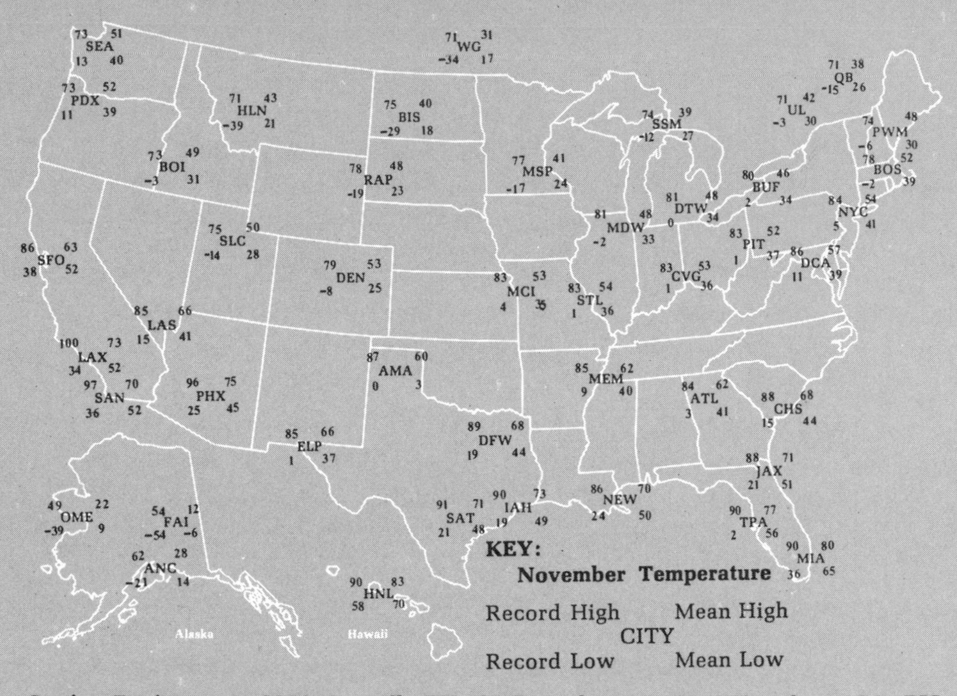

Station Designators: AMA Amarillo TX; **ANC** Anchorage AK; **ATL** Atlanta GA; **BIS** Bismarck ND; **BOI** Boise ID; **BOS** Boston MA; **BUF** Buffalo NY; **CHS** Charleston SC; **CVG** Cincinnati OH; **DCA** Washington DC; **DEN** Denver CO; **DFW** Dallas-Fort Worth TX; **DTW** Detroit MI; **ELP** El Paso TX; **FAI** Fairbanks AK; **HLN** Helena MT; **HNL** Honolulu HI; **IAH** Houston TX; **JAX** Jacksonville FL; **LAS** Las Vegas NV; **LAX** Los Angeles CA; **MCI** Kansas City MO; **MDW** Chicago IL; **MEM** Memphis TN; **MIA** Miami FL; **MSP** Minneapolis-St. Paul MN; **NEW** New Orleans LA; **NYC** New York NY; **OME** Nome AK; **PDX** Portland OR; **PHX** Phoenix AZ; **PIT** Pittsburgh PA; **PWN** Portland ME; **QB** Quebec QUE; **RAP** Rapid City SD; **SAN** San Diego CA; **SAT** San Antonio TX; **SEA** Seattle WA; **SFO** San Francisco CA; **SLC** Salt Lake City UT; **SSM** Sault Ste. Marie MI; **STL** St. Louis MO; **TPA** Tampa FL; **UL** Montreal QUE; **WG** Winnipeg MAN; **YC** Calgary ALB.

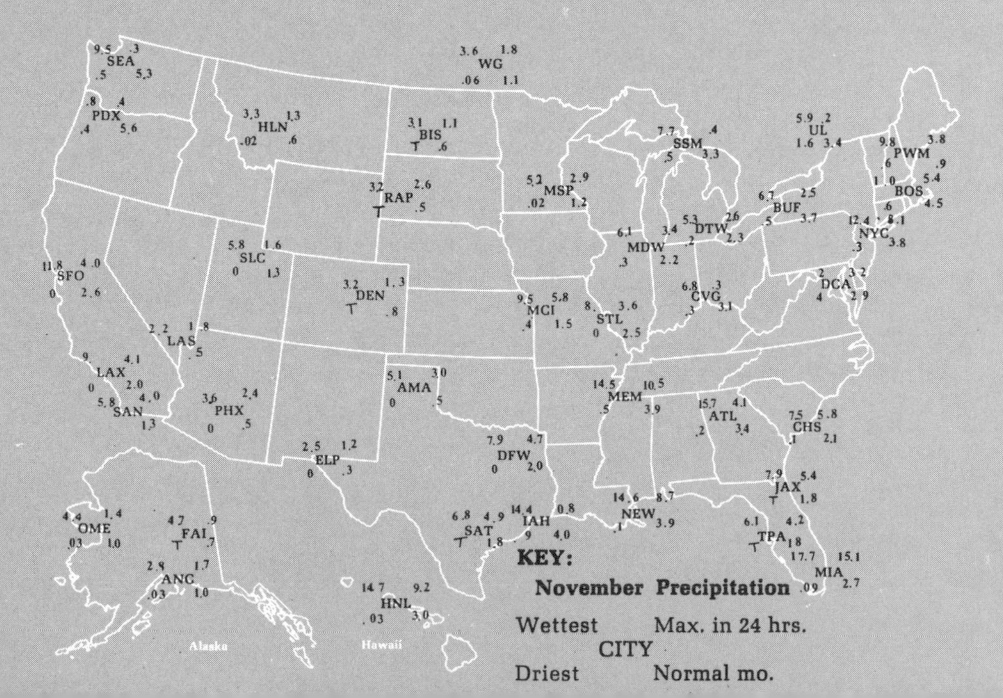

Fog

The fog comes
on little cat feet.
It sits looking
over the harbor and city
on silent haunches
and then moves on.
— Carl Sandburg, "Fog"

The most quoted description of fog ever written was the first sentence of Sandburg's little poem. In a less poetic vein, the eminent American climatologist Robert DeCourcy Ward introduced the subject with another succinct description in his *Climates of the United States* (1925):

> Fog is silent in its formation and disappearance; unaccompanied by violent atmospheric disturbances of any sort; associated with no brilliant optical phenomena; suggesting no startling worldwide human responses, fog nevertheless has many important relations to the life of man.

Fog is a cloud form at the surface of the earth consisting of a multitude of minute water droplets suspended in the atmosphere. According to international definition, fog is present when horizontal visibility is reduced below one kilometer (0.62 mi). Heavy fog occurs when the visibility is 0.4 km (0.25 mi) or less.

Fogs of all types originate when the dew point and the air temperature become identical, or nearly so, provided that sufficient condensation nuclei are available in the air. This may occur either through cooling the air to its dew point (producing advective fog, radiation fog, or upslope fog), or by adding moisture and thereby elevating the dew point (producing steam fog or frontal fog). Fog seldom tends to form when the dew point temperature differs from air temperature by more than 4F (2C) degrees.

Foggiest in the United States, West Coast — Cape Disappointment, Washington, at the mouth of the Columbia River, averages 2552 hours (equal to 106 complete days) of heavy fog annually. The foggiest period is normally August and September, with a decrease during fall and winter to a minimum in April and May. In summer, Pacific coastal locations have almost daily fog that burns off in late morning only to return in late afternoon.

At Eureka, in northwest California, heavy fog is observed at some time on 49 days of the year; and the airport at nearby Arcata is considered the most suitable in the nation to test fog-dispersal apparatus. San Francisco Airport, on the bay, has 18 days with heavy fog. In the Los Angeles area, the downtown Civic Center district has 17 days per year with heavy fog, while the airport ten miles (16 km) to the southwest, very near the seacoast, has 47 days. San Diego Airport, fronting the bay, has 29 heavy fog days.

In the mountains in the north, Stampede Pass, Washington, at 3958 ft (1203 m), has heavy fog 252 days annually at some hourly observations; a maximum is observed in December and a minimum in September. In the Sierra, Blue Canyon, a mile high at 5280 ft (1609 m), has 68 days with heavy fog, practically all occurring in the winter season when cyclonic activity is present.

Foggiest in the United States, East Coast — The rockbound coast of Maine has the highest number of hours with fog on the Atlantic Coast of the United States. Moose Peak Lighthouse on Mistake Island, at an elevation of 72 ft (22 m), averages 1580 hours per year of heavy fog. Mistake Island is located about halfway between Mt. Desert and Eastport, Maine, in the extreme northeast corner of the country. The latter has 65 days with some heavy fog observed, while Portland, farther south, reports 55 days. Logan International Airport, bordering Boston Harbor, averages only 23 days, but Nantucket Island, with full sea exposure, experiences 85 days with fog at some time of day. The foggiest period comes during summer. Mt. Washington at 6262 ft (1909 m), in the clouds most of the time, reports 308 days with fog evenly distributed throughout the year. At the other extreme, Key West, Florida, averages less than one day per year with heavy fog.

Foggiest Inland — The windward slopes of the mountain chains have the greatest frequency of heavy fog, especially when the prevailing wind is from a moisture source. The higher elevations of the Appalachian Mountains show a peak in West Virginia, with over 100 days of fog per year. Elkins, at an elevation of 1948 ft (594 m), has 81 days with heavy fog. The slopes of the eastern foothills of the Rocky Mountains have a higher frequency of fog than do the Great Plains to the east. The orographic effect of the rising terrain increases the number of days with fog as one progresses westward. Cheyenne, in the southeast corner of Wyoming, has 23 days with heavy fog. The region around the Great Lakes is much foggier than at a distance from the water. Lake Superior has a maximum with as many as 53 days at Duluth, Minnesota, while St. Louis has only 10 days with heavy fog. From the Sierra to the Rocky Mountains the lack of moisture results in a low fog frequency. No station averages over 10 days with heavy fog. But in the Northwest from the Cascades of Washington and Oregon to the Rocky Mountains, all stations show 11 days or more of heavy fog, and two have as many as 30 days.

Unusual Accidents on Land Resulting from Fog

1969 November 29 — Chain-reaction crash of 21 vehicles occurred in fog at 8:00 A.M. on New Jersey Turnpike, one mile (1.6 km) north of Exit No. 2, about halfway from Camden to the Delaware Memorial Bridge. Propane truck jacknifed, was struck by a following trailer truck, and others piled into the fiery mass. Six persons were killed and at least 17 others injured.

1969 December 15 — Multiple collisions involved over 100 vehicles on a fog-shrouded, 12-mile (19.2-km) stretch of Santa Ana Freeway in Orange County, southeast of Los Angeles. Most were rear-end crashes.

1973 October 24—Sixty-five vehicles crashed on the New Jersey Turnpike, near Interchange 16W close to the Meadowlands Stadium. Nine persons were killed and 40 injured. Smoke from a fire a mile away mixed with fog to create zero visibility.

1980 November 10 — As many as two dozen vehicles, including eight tractor-trailers, were involved in multiple accidents along 500 yards of Interstate 15, near San Bernardino, California. At least seven persons were killed and 17 injured in a fiery collision on the fog-shrouded freeway. Cars descending a long, fog-free grade from Cajon Pass suddenly entered dense fog with zero visibility.

Overleaf: Fog blown down the Cape out to the ocean. Marshfield, Massachusetts. Boston Herald photo.

Unusual Accidents on Water Resulting from Fog

1852 August 20 — S.S. *Atlantic* eastbound was rammed on Lake Erie by S.S. *Ogdensburg* westbound in dense fog about 6 mi (10 km) off Long Point in southeast Ontario. There were 500 persons aboard *Atlantic* and an unknown number on *Ogdensburg*. A total of 250 persons, many of them Norwegian immigrants, were drowned.

1901 February 22 — Pacific Mail steamer *City of Rio de Janeiro* struck a rock while entering the Golden Gate to San Francisco Bay in a dense fog and sank in a few minutes. Fog had lifted momentarily, permitting ship to attempt entrance, but then settled in quickly again with a northeast wind on the bow. A total of 128 persons, mostly Japanese and Chinese immigrants, were lost.

1906 November 22 — *Valencia*, northbound from San Francisco to Puget Sound ports, grounded on Vancouver Island, British Columbia, "during thick fog" and southeast storm, resulting in the loss of 129 persons.

1914 May 29 — Canadian Pacific liner *Empress of Ireland* (14,000 tons; 8 years old) was rammed and sunk in 15 minutes by small Norwegian collier *Storstad* in St. Lawrence River off Ste. Luce, about 180 mi (290 km) seaward from Quebec City. Shallow river fog obscured the vision of both vessels shortly before the crash, though each had sighted the other at a distance. There were 1024 fatalities of a total of 1477 aboard.

1921 August 6 — Steamer *Alaska*, Seattle to San Francisco, ran on Blunt's Reef in dense fog just off Cape Mendocino, California. Forty-nine lives were lost.

1923 September 9 — Seven U.S. Navy World War I four-stack destroyers, running in close formation at 20 knots (37 km/h), struck the rocks off Point Arguella, California, in dense fog. Faulty use of navigation information and excessive speed in fog were blamed for the accident. All ships were total wrecks and 22 men died.

1929 August 30 — Coastal steamer *San Juan* (2152 tons) was rammed off Pigeon Point near Santa Cruz, California, by oil tanker S.C.T. *Dodd*, which was proceeding at "excessive speed in fog without sounding fog signals." Seventy passengers and crew of *San Juan* drowned.

1951 April 20 — ESSO tanker *Suez*, moving at full speed of 18 knots through dense fog, rammed ESSO tanker *Greensboro* about 200 mi (322 km) south of Morgan City, Louisiana, in the Gulf of Mexico. Explosion and fire on *Greensboro* resulted in loss of 37 of crew of 42. Two were lost on *Suez*. *Greensboro* was towed into port with 100,000 barrels of oil still in its tanks.

1956 July 25 — Italian luxury liner *Andrea Doria* (29,000 tons) was rammed by Swedish liner *Stockholm* (12,644 tons) near Nantucket Lightship. "The night was mild with thick fog," according to the press. *Stockholm* was moving at a normal speed of 18 knots "in clear moonlight with stars visible," Europe-bound, while *Andrea Doria* was approaching the United States coast from the east through fog. As a result of radar surveillance, each ship was aware of the other's presence, and both were making confused avoidance maneuvers when *Andrea Doria* emerged from the fog across the bow of *Stockholm*. The ships crashed and then backed away. *Andrea Doria* soon lost her stability and slowly sank, making the final plunge 12 hours after the accident. The human toll was 52, either killed by the impact of the crash or drowned before or during the rescue operations.

1964 November 26 — Israeli liner *Shalom* collided with Norwegian tanker *Stolt Dagali* off New Jersey coast in dense fog. Nineteen of the tanker's crew were lost.

December

Barnaby bright, Barnaby bright,
The longest day and the shortest night;
Lucy light, Lucy light,
The shortest day and the longest night.

— Old English rhyme, referring to St. Barnabas' Day (the summer solstice) and St. Lucy's Day (the winter solstice)

What freezings have I felt, what dark days seen!
What old December's bareness every where.

— Shakespeare, Sonnet 97

Opposite: Ice glaze up to three inches thick coats limbs during severe ice storm, Springfield, Illinois, March 26, 1978. State Journal-Register photo by William Hagen.

The sun that brief December day / Rose cheerless over hills of gray, / And, darkly circled, gave at noon / A sadder light than waning moon." The opening lines of Whittier's *Snow-Bound* well express the setting of many December days when a storm situation is developing. The sun continues to sink lower and lower during the first three weeks of the month, until its direct rays fall on the Tropic of Capricorn, very near São Paulo, Brazil, and Antofagasta, Chile. At the time of the winter solstice, on or about December 21, the sun at noon will stand only 42° above the horizon at Key West and only 28° above the horizon at the 49th parallel, the border between the western states and Canada. Shakespeare referred to "dark December" in *Cymbeline.* Not only is the duration of daylight at its shortest now, but cloudiness is at its annual maximum since the continent is almost circled by storm-generating areas. In the northern Great Lakes region the sun will shine only about 25 percent of the time, and along the Northwest coast of Washington the amount of sunshine drops to about 15 percent of the possible on a December day.

December is also the stormiest month of the year for the United States as a whole. Its last week has produced some notable weather events: New York City's deepest snowstorm came on December 26, 1947; New England's coldest temperature of −50°F (−46°C) was registered in Vermont on December 30, 1933; the coldest air mass ever to come out of central Canada into the Midwest crossed the border on December 30–31, 1864, and dropped the thermometer at Chicago to an early record of −25°F (−32°C); and Southern California experienced its worst storm situation on December 31, 1933, when Los Angeles recorded 7.36 in (187 mm) of rain in twenty-four hours.

The Pacific High remains in the same relative position as in November, centered about 1200 mi (1900 km) west-southwest of San Diego; it is joined by a broad band of high pressure to a strong anticyclonic center over the Intermountain region of Idaho, Utah, and Nevada. The Pacific High is responsible for the frequent periods of fine weather that make winters in California a pleasant and tolerable season. The Aleutian Low, exhibiting its greatest strength of the year, has a principal center over the western Aleutians, where mean pressure reads 29.53 in (100.0 kPa), and a secondary center covers the northwest quarter of the Gulf of Alaska. Activity reaches the year's maximum in this region, and storms are dispatched southeast at frequent intervals to batter the west coast of Canada and the United States.

In the Atlantic Ocean, the Azores–Bermuda High holds to the same latitude as in November, but the center has shifted well to the east to a position off the coast of northwest Africa. The Iceland–Greenland Low becomes the main feature of North Atlantic weather maps, with a single center located southwest of Iceland and southeast of Greenland. Mean pressure is down to 29.46 in (99.8 kPa), indicating a stormy time for the shipping and air lanes not far to the south.

The main Pacific jet stream follows a pattern similar to November's except that its path from the Ohio Valley continues directly east to leave the continent over Virginia. A new jet stream from the central Pacific Ocean appears this month over the southern half of the country; entering over Baja California, it proceeds east to central Texas, then swings northeast to the Carolina Capes. The increased storm activity in Texas and the Gulf of Mexico is stimulated by the southern jet bringing cold and warm airstreams into close proximity.

Dark December is the period of great cyclonic storms. The Gulf of Alaska harbors more storm activity than any other part of the Northern Hemisphere does at any time of the year. Another region experiencing its greatest storm frequency is the Great Lakes. Cyclonic formation in the Gulf of Mexico, too, reaches a maximum in December.

The principal storm tracks move farther south than in November. The main path of Pacific and Alberta lows crosses the southern Prairies and swings north of Lake Superior, traversing Ontario and southern Quebec before leaving the continent over southern Labrador. The now well-frequented track from the Gulf of Mexico follows the Atlantic seaboard to Cape Hatteras, whence an ocean passage carries to Newfoundland and an ultimate rendezvous with the Iceland–Greenland Low.

By December the last vestiges of the autumn anticyclone pattern have disappeared and certain features unique to winter appear, such as a southerly track across the Pacific Ocean located near 30°N, occasional centers of high pressure in southern Quebec, and a secondary path of polar highs from the Dakotas south to Texas. An outstanding feature of December is the continued intensification of anticyclonic activity in the Intermountain region, where the number of December high-pressure centers exceeds that of any other region. A close relationship exists between the Intermountain high pressure and the low pressure in the Gulf of Alaska. When the pressure is high over the western United States, storms from the Alaskan region are shunted to the north and enter the continent over northern British Columbia; when the pressure is low, they enter at lower and lower latitudes until the last of the series may come over California.

The winter precipitation pattern is well established by December with two regions of heavy rainfall: one in the interior Gulf States, the other on the coasts of Washington and Oregon. The 4.0-in (102-mm) area covers all the Gulf Coast from east Texas to northwest Florida and extends inland over most of Arkansas and Tennessee. The Florida peninsula is quite dry, with central and southwest portions receiving less than 2.0 in (51 mm). The Great Plains experience a dry regime also; the one-inch (25-mm) line runs from near Duluth, Minnesota, south-southwest through western Iowa and eastern Kansas to central Texas. Precipitation is at or close to the annual maximum for any month in the Northwest, where 8.0 in (203 mm) fall at many valley locations and above 16 in (406 mm) at high elevations. The Sierra of California receive over 8.0 in (203 mm) and the mountains behind Los Angeles have 4.0 in (102 mm).

December also exhibits the winter temperature regime, with mean readings below 10°F (−12°C) in North Dakota and above 60°F (16°C) in south Florida and southeast Texas. The 50°F (10°C) isotherm runs from South Carolina west to northern Louisiana and on to the Great Bend of the Rio Grande; and the 40°F (4°C) line loops out from Chesapeake Bay around the southern Appalachians to Tennessee, then runs through Arkansas to the southeast corner of New Mexico. Mean freezing follows a line from near Boston southwestward into West Virginia, then north to Cleveland and west to near Cheyenne, Wyoming. Parts of northern New England, northern New York, and the northern Plains are within the less-than-20°F (−7°C) area. Temperatures along the Northwest coast average above 40°F (4°C); farther south in California the means are generally above 50°F (10°C) from north of San Francisco to the Mexican border.

Temperature extremes for December: maximum 100°F (38°C) at La Mesa, California, December 8, 1938; minimum −59°F (−51°C) at West Yellowstone, Montana, December 19, 1924.

Throughout the central and northern sections of the country there is, as a rule, little farm activity during December. The marketing of farm products continues active, and, when weather permits, some shock corn husking is accomplished. Much citrus fruit is usually gathered in Florida, and the harvesting of oranges, olives, and winter truck goes forward in California. This is a critical frost month in the citrus districts of Florida, Texas, Arizona, and California.

December

1 **1831** Coldest December ever closed Erie Canal for entire month; New York City had a record low mean of 22°F (−6°C).

2 **1925** Late-season hurricane moved off Georgia, later crossed Cape Hatteras as tropical storm.

1950 Late-season tornado killed four in Madison and Bond counties in Illinois.

3 **1791** Salem, Mass.: "Nivis tempestates per totum diem" (Rev. Wm. Bentley's diary).

1856 Severe blizzard-type storm raged for three days in Kansas and Iowa; early pioneers suffered.

4 **1786** First of two great early December snows began; 18 in (46 cm) at Morristown, N.J., 20 in (51 cm) at New Haven, Conn.; high tide at Nantucket did great damage.

5 **1784** "Great Winter Freshet" occurred on Merrimack River in New Hampshire, highest since 1740; high flood also on Connecticut River at Hartford.

1953 Killer tornado at Vicksburg, Miss.; 38 dead, 270 injured; $25 million damage.

6 **1886** Great Southern Appalachian Snowstorm; 42 in (107 cm) reported in mountains from three-day storm; 33 in (84 cm) at Asheville, N.C.; 25 in (64 cm) at Rome, Ga.

7 1740 Greatest flood in Connecticut in 50 years, on Merrimack in 70 years, when early winter broke up during heavy rainstorm.

8 1876 Term *blizzard* first employed in government's *Monthly Weather Review.*

1938 La Mesa, Calif., registered 100°F (38°C), warmest ever in U.S. during December.

9 1786 Second great snowstorm in five days; 21 in (53 cm) at Morristown, N.J., 17 in (43 cm) at New Haven, Conn.; "four feet on level" in eastern Massachusetts.

10 1946 New York City hit 70°F (21°C), warmest ever in December.

11 1932 San Francisco's coldest day with minimum of 27°F (−3°C) and maximum of 35°F (2°C); airport dropped to 20°F (−7°C); 0.8-in (2-cm) snowfall.

12 1960 Pre-Winter Blizzard in Northeast; 20.4 in (53 cm) at Newark, N.J.; Nantucket had 15.7 in (40 cm) with wind averaging 35.8 mi/h (58 km/h) with gusts to 51 mi/h (82 km/h).

13 1962 Severe Florida freeze, coldest of twentieth century in December; Jacksonville 12°F (−11°C); Tampa 18°F (−8°C); Miami 35°F (2°C); millions in damage to crops and foliage.

14 1924 Spectacular temperature drop at Helena, Mont.; fell 79°F (44°C) in 24 hours, 88°F (49°C) in 34 hours on December 14–15, from 63°F to −25°F (17°C to −32°C).

15 1839 First of Triple December Storms on Massachusetts Bay; great loss of life at Gloucester in high gales; over 20-in (61-cm) snowfall interior New England.

16 1835 Cold Wednesday, New England's bitterest daylight; at noon Hanover, N.H., −17°F (−27°C), Boston −4°F (−20°C); northwest gales all day; great New York City fire in night destroyed much of financial district.

1917 Ice jam formed on Ohio River between Warsaw, Ky., and Rising Sun, Ind.; held for 58 days; ice 30 ft (9 m) high backed river up for 100 mi (161 km).

17 1924 Severe ice storm in central Illinois; 3.63 in (92 mm) of precipitation froze; 1.9 in (4.8 cm) of sleet and ice on ground for two weeks; ice on 12 in (30 cm) of wire weighed 8 oz (227 gm).

18 1957 Tornado in southern Illinois (Jackson, Williamson, and Franklin counties) killed 11 people.

19 1777 Continental Army moved into encampment at Valley Forge on day having "stormy winds and piercing cold"; moderate in-and-out winter followed.

1924 Riverside Ranger Station in Yellowstone Park dropped to −59°F (−51°C), lowest ever in conterminous 48 states in December.

20 1836 Famous "Sudden Change" in central Illinois; cold front at noon dropped temperature 40°F (22°C) to 0°F (−18°C) in minutes; "chickens frozen in tracks . . . men frozen to saddles" and other tall tales.

21 1964 Great warm surge from Pacific over Northern California and Oregon brought torrential rains on deep snow cover, resulting in record floods.

22 1839 Second of Triple December Storms; 25 in (64 cm) snow at Gettysburg; gales in New England but only light snow on coast.

23 1775 The "Snow Campaign" in Carolinas and Georgia; 24 in (61 cm) fell at American camp at Reedy River, N.C.

1811 Cold Storm on Long Island Sound; 0°F (−18°C) cold, 12 in (30 cm) snow; ships wrecked and crews frozen.

24 1872 Extreme cold in Midwest; Chicago set a record (1871–1981) at −23°F (−31°C); Minneapolis rose from morning low of −38°F (−39°C) only to −17°F (−27°C) at 2:00 P.M.

25 1980 "Cold Christmas of '80" in Northeast; sharp cold front moved southeast during daylight hours; Old Forge, N.Y. −38°F (−30°C); Boston dropped from midnight 32°F (0°C) to −7°F (−22°C); New York City from 37°F (3°C) to −1°F (−18°C).

1778 The "Hessian Storm" began at Newport, R.I., where many British mercenaries froze to death; deep snows, gales in all New England; Cambridge −8°F (−22°C), for coldest Christmas.

26 1776 Washington crossed ice-clogged Delaware River, marched on Trenton in snow and sleet storm; surprised and captured many of Hessian garrison.

1947 New York City's deepest snowstorm; 26.4 in (67 cm) in Central Park in 24 hours; 32 in (81 cm) in suburbs; traffic completely stopped; removal cost $8 million; 27 died.

27 1969 Big Post-Christmas Snowstorm in New York and Vermont; record storm totals of 39 in (99 cm) at Montpelier and 29.8 in (76 cm) at Burlington; moderate temperatures; public emergency declared in Vermont.

28 1839 Third of Triple Storms; 24 in (61 cm) more snow in Hartford–Worcester area; Boston barometer 28.77 in (97.4 kPa); whole gales swept coast causing many wrecks.

1924 Coldest December morning ever in Iowa; 104 stations averaged −24.6°F (−31.4°C) minimum.

29 1830 Very heavy snow: Kansas City 36 in (91 cm) and Peoria 30 in (76 cm); beginning of famous "Winter of the Deep Snow," which continued until mid-February; great suffering among pioneers.

1894 Severe Florida freeze; fruit destroyed; considerable tree damage.

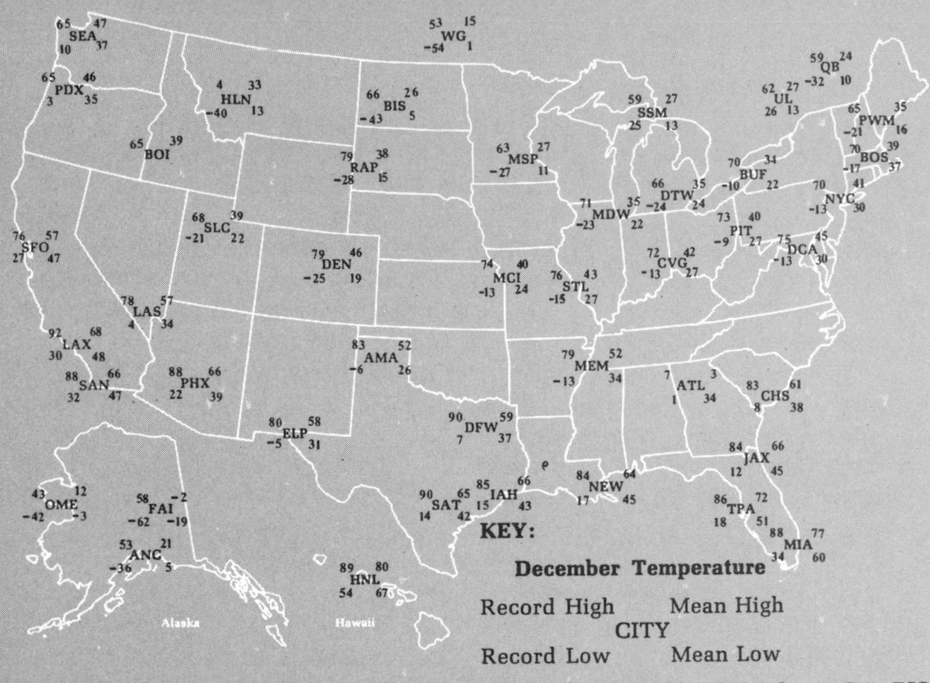

Station Designators: AMA Amarillo TX; **ANC** Anchorage AK; **ATL** Atlanta GA; **BIS** Bismarck ND; **BOI** Boise ID; **BOS** Boston MA; **BUF** Buffalo NY; **CHS** Charleston SC; **CVG** Cincinnati OH; **DCA** Washington DC; **DEN** Denver CO; **DFW** Dallas-Fort Worth TX; **DTW** Detroit MI; **ELP** El Paso TX; **FAI** Fairbanks AK; **HLN** Helena MT; **HNL** Honolulu HI; **IAH** Houston TX; **JAX** Jacksonville FL; **LAS** Las Vegas NV; **LAX** Los Angeles CA; **MCI** Kansas City MO; **MDW** Chicago IL; **MEM** Memphis TN; **MIA** Miami FL; **MSP** Minneapolis-St. Paul MN; **NEW** New Orleans LA; **NYC** New York NY; **OME** Nome AK; **PDX** Portland OR; **PHX** Phoenix AZ; **PIT** Pittsburgh PA; **PWN** Portland ME; **QB** Quebec QUE; **RAP** Rapid City SD; **SAN** San Diego CA; **SAT** San Antonio TX; **SEA** Seattle WA; **SFO** San Francisco CA; **SLC** Salt Lake City UT; **SSM** Sault Ste. Marie MI; **STL** St. Louis MO; **TPA** Tampa FL; **UL** Montreal QUE; **WG** Winnipeg MAN; **YC** Calgary ALB.

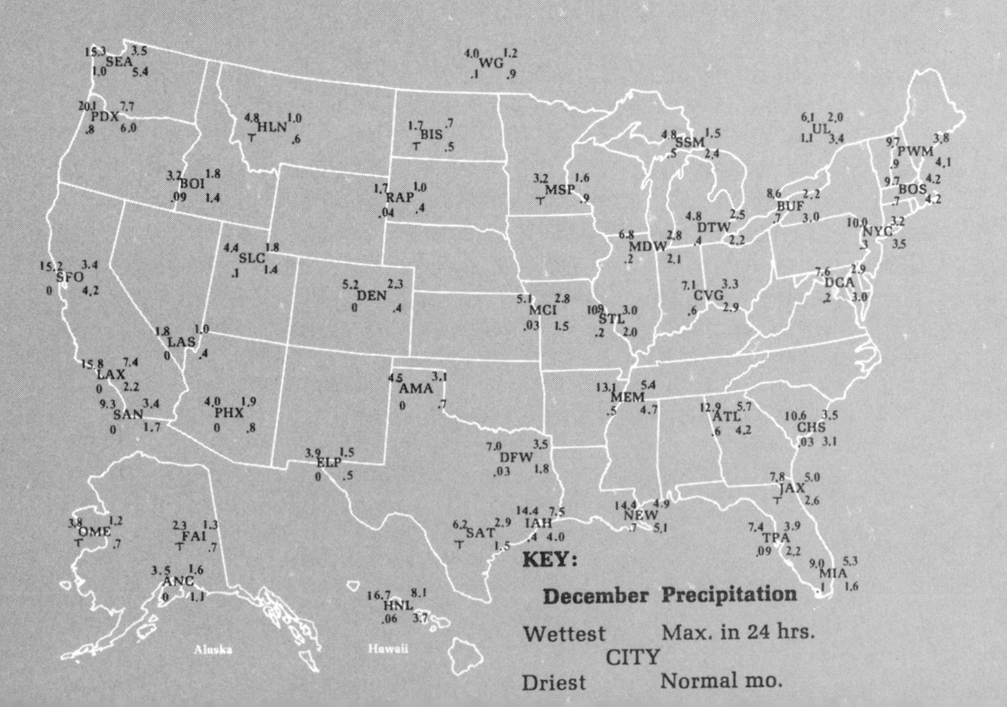

1919 American Meteorological Society founded at St. Louis.

30 **1917** Great cold wave set many December records in Northeast; New York City −13°F (−25°C), Boston −15°F (−26°C) with below zero on five nights; −44°F (−42°C) in New Hampshire.

1933 Coldest in modern records for New England: −50°F (−46°C) at Bloomfield, in Vermont's Northeast Kingdom.

31 **1775** Americans under Montgomery and Arnold assaulted Quebec during snowstorm in early morning; attack ended in disaster.

1933 Los Angeles had 24-hour rain of 7.36 in (187 mm), setting stage for area's worst flood.

1963 Great New Year's Eve Snowstorm in Deep South; Meridian, Miss., 15 in (38 cm), Bay St. Louis on Mississippi Gulf Coast 10.5 in (27 cm), New Orleans 4.5 in (11 cm); freeze followed on New Year's Day; traffic to Sugar Bowl hampered.

Ice Storms

He giveth forth snow like wool; he scattereth the hoarfrost like ashes.
He casteth forth his ice like morsels: who can stand before his cold?

—Psalms 147: 16, 17

Winter's worst menace in modern times is a freezing rain that coats all outside objects with an icy sheath known to meteorologists as glaze. With our almost complete dependence on electric power for heating, cooking, communications, and even entertainment, the disruption attending an ice storm can cause widespread hardship and malaise. One may escape the effects of a big snowstorm by seeking shelter, but the chill and discomfort of an ice storm can make themselves felt inside as well as out.

Freezing rain occurs when water droplets fall from an above-freezing layer of air aloft through a shallow layer of below-freezing air at the surface of the earth. Upon impact or shortly thereafter the droplets freeze on all exposed objects, coating them with a varying thickness of glaze. If the layer of cold air at the surface is deep, the rain droplets will freeze in their descent and form small ice pellets, known as sleet. Sometimes the two forms will occur at different times during one storm, congealing all into an icy mass covering the ground and all exposed objects. An ice storm is popularly known as a "silver thaw" in some sections. Despite the damage and inconvenience caused by an ice storm, it presents a most spectacular scene when the rays of the rising sun glint with thousands of beautiful spangles on ice-coated trees and shrubs.

Region — The greatest prevalence of winter icing extends in a broad band from the south-central Great Plains northeast to the Great Lakes and east to the Middle Atlantic States and New England. Freezing rain usually occurs in a narrow band 50 to 100 mi (80 to 161 km) wide to the northwest of the track of cyclonic storms moving northeast. Often alternate zones of rain, freezing rain, ice pellets (sleet), and snow extend outward from the storm center. Certain elevations in mountainous sections, when horizontal temperature zones are properly aligned, are also subject to icing conditions, as are deep valleys where cold air is trapped below warmer air aloft. Severe ice storms have occurred in the South, too, especially across the northern parts of the Gulf States, but even such southerly locations as Jacksonville, Florida, and Brownsville, Texas, have on rare occasions experienced light icing. The western mountains are also subject to icing, and some of the most spectacular ice storms have occurred in the gorge country of the lower Columbia Valley when the weather forces are properly marshaled with alternate strata of cold and warm air struggling for dominance of that strategic location.

Amount — Ice forming on exposed objects generally ranges from a very thin coating to about an inch (2.5 cm) thick. But deposits up to 8.0 in (20 cm) were reported in northern Idaho January 1–3, 1961, up to 6.0 in (15 cm) in northwest Texas November 22–24, 1940, and to 6.0 in in New York State December 29–30, 1942. Sometimes wet snow following an icing increases the clinging accumulations to fantastic sizes. Utility poles are often broken down by such burdens; and many radio towers have bent double under their icy load.

An ice coating on a length of No. 14 telephone wire, 12 in (30 cm) long, weighed 11 lb (5 kg) in Michigan's famous ice storm of February 21–23, 1922. A loading of 4.5 lb (2 kg) per foot (0.3 m) was reported in Illinois on January 1, 1948. During the memorable late November 1921 storm at Worcester, Massachusetts, an evergreen tree, 50 ft (15 m) high with an average spread of 20 ft (6 m), was estimated to have a burden of ice weighing five tons (5.5 metric tons).

Most Damaging — The greatest ice storm in United States history occurred over a wide area from Texas to West Virginia from January 28 to February 1, 1951. Total damages were estimated at nearly $100 million. The severest blows fell in a zone in the north Gulf States and across Tennessee. Ice formed on exposed objects to a thickness of 0.5 to 4.0 in (1 to 10 cm) and accumulated on the warm ground as a partly congealed cover. Snowfalls of about 4.0 in (10 cm) fell on top of this. Then an invasion of arctic air after the storm froze all of this into a solid mass that could not be easily removed from the streets. Traffic in the Upper South was impeded for as long as ten days afterward.

The Nashville Tennessean described the situation: "Never before in recorded history did winter hit this area with such devastating force to paralyze the everyday life of the community. Never before were so many homes plunged into frigid darkness. Never before were so many faced with hardship and suffering in what was once the security of home, and never before were the complex transportation systems of the area held immovable for such an extended period."

The total damage of nearly $100 million broke down as follows: forests $56.0 million, communication and power lines $10, highways $15, fruit trees $4, buildings and plumbing $4.3, livestock $3, and crops $1.6. About 25 persons lost their lives as a result of accidents and over 500 were seriously injured in mishaps on the ice.

Some Severe Ice Storms

New England — Outstanding ice storms hit interior New England in February 1659, December 1699, January 1788, March 1837, January 1886, November 1921, January 1953, February 1959, December 1969, and December 1973.

The effects of an ice storm in colonial days were related in the diary of Judge Samuel Sewall of Boston:

> Fifth-day, Nov the last. 1699. The Rain freezes upon the branches of the Trees to that thickness and weight, that great havock is thereby made of the Wood and Timber. Many young and strong Trees are broken off in the midst; and multitudes of Boughs rent off. Considerable hurt is done in Orchards. Two of our Apple-trees are broken down, Unkles Tree, two-thirds of it, are broken down. Peach Trees at Mrs. Moodeys are almost all spoil'd. And my little Cedar almost quite mortified. Some think the Spoil that is made amounts to Thousands of pounds. How suddenly and with surprise can God destroy!

The worst ice storm of this century in central New England affected a strip from northern Connecticut to southern Maine from November 26 to 29, 1921. After a day and a half of alternate snow, sleet, and rain, the latter continued even though the surface temperature fell to 25°F (−4°C). Professor Charles F. Brooks, the founder of the American Meteorological Society, described the scene near his home at Worcester, Massachusetts, the center of the heaviest icing:

> At night it began to rain steadily and ice as steadily to form. Morning saw an inch-thick armor of ice over everything out-of-doors, and still it rained and froze, while the northeast wind increased to a gale. By afternoon city streets were dangerous. Everywhere branches and trees and electric wires and poles were falling. A wild and terrible night followed. Electric lights were extinguished, and cities and towns lay in blackness, trolley cars ceased running, telegraph and telephone service was gone, streets were impassable to vehicles, some of them to pedestrians. The climax was a thunderstorm, the thunder crashing to an accompaniment of falling trees. (*Bulletin of the American Meteorological Society.*)

Ice-coated electric wires reached 3 in (7.6 cm) in diameter and weighed nearly a pound (0.5 kg) per foot (0.3 m).

One does not need to check back further than December 1973 for a chilling example of the disruption in community life that a steady freezing rain of twenty-four hours' duration can bring. Most of southern New England received over an inch (25 mm) of freezing precipitation on December 16–17, 1973, which

Opposite: Tree damage at Springfield, Illinois, after ice storm, January 20–21, 1959. State Journal-Register photo.

coated wires and exposed objects with an icy sheath 0.5 in (1.3 cm) thick. The greatest power outages in New England history resulted with 250,000 customers without electricity in Connecticut alone. Tree damage in the Nutmeg State was said to have been greater than in the Hurricane of 1938. Many homes remained without power or telephone for a week after the storm. This storm also struck devastating blows on central Long Island.

Northeast — Freezing rain fell across Pennsylvania, New Jersey, and New York on January 8 to 11, 1953. In the Keystone State after 40 hours of ice accumulations there were generally 1 to 2.5 in (2.5 to 6 cm) of ice, with extreme reports up to 4.37 in (11 cm). A wire 15 in (38 cm) long weighed nearly 2 lb (0.9 kg) and circumferences reached 10 to 13 in (25 to 33 cm). Fifty thousand customers were without electricity and telephone service in Pennsylvania. In New Jersey, seventy thousand homes were without power, and utility damages were placed at $1 million. In southeast New York, where ice accumulations were 2 to 3 in (5 to 8 cm) thick, 75 main power lines were broken in Westchester County alone and many homes were without power for 72 hours.

Illinois — Until Chicago's Big Snow of '67 came long, the state's most damaging winter storm was a freezing rain on December 17–18, 1924. More than $21 million in damage resulted in all sections of the state except the extreme south. As usual, the worst destruction was in the central portions, where both sleet and glaze were experienced. Springfield had 3.83 in (92 mm) of mixed precipitation, with extensive tree damage resulting. An ice-coated twig weighed 15 times its normal weight. The depth of sleet and ice on the ground was 1.9 in (5 cm). Ice remained on trees until January 4. Some electric power was not restored until January 10. The first long distance telephone call went through six days after the storm, and the streetcar system in Springfield was not back to normal until January 4. Another severe ice storm visited the same Springfield area and central Illinois on the weekend of March 25–26, 1978, causing widespread blackouts and disrupted communications.

Chicago's worst ice experience came on January 1, 1948, when

The Cunard Royal Mail steamship Britannia *leaves Boston on February 3, 1844, through a seven-mile path cut in ice by local port authorities.*

ice and snow built up on wires to a thickness of 2.0 in (5 cm). Winds downtown gusted to 50 mi/h (80 km/h). Radio station antennas suffered, with eleven destroyed or badly damaged in the Chicago area alone. It was the most disastrous storm ever experienced by local utilities companies; damage in Illinois amounted to $3 million. It was also severe in eastern Iowa and southern Wisconsin.

Wisconsin-Michigan — Oldtimers still refer to February 21–23, 1922, for their worst icing experience. A belt from 50 to 100 mi (80 to 161 km) wide, west to east, received over 4.0 in (104 mm) of precipitation in a 24-hour period. The accumulations added 20 to 30 times the normal weight to branches and trees. The lower Fox River Valley in Wisconsin suffered severely. The damage area in southern Michigan lay just north of a Muskegon–Bay City line.

Losses to utilities in Wisconsin amounted to $10 million, and in Michigan to $5 million.

Atlanta Area — Northern Georgia lies in a belt of occasional severe icing when a storm passes to the southeast in wintertime. The area's greatest ice storm covered a 60-mi (97-km) wide belt across Alabama and Georgia during December 27–29, 1935. All traffic was obstructed for several days. Over 25,000 homes were without electric power, and damage to utility companies amounted to $1 million. An ice mass covered the ground to a depth of 2 in (5 cm), and a sheath of ice 1.75 in (4 cm) thick encased wires and trees. Light winds saved fruit trees from major damage. With the weather remaining cold, the ice clung to objects until January 2.

More recently, in 1973, the Atlanta area experienced a 4.09-in (104-mm) rainfall on January 7 to 9 with temperatures at or below freezing. Traffic was paralyzed for two days, and communications hampered for a week by the combination of ice and snow covering the ground and clinging to exposed objects. Damage to utilities and to trees reached an estimated $20 million.

Fort Worth–Dallas Area — Severe icing, "the worst on record," was experienced on January 6–9, 1937. As much as 2.0 in (5 cm) formed on wires. Communications were disrupted and highway traffic made extremely hazardous. Total damage was placed at $3 to 4 million. Ice did not disappear until January 12.

The Pacific Northwest — With a mixed snow-sleet-glaze precipitation falling, the Columbia Gorge Storm filled the narrow gap in the Cascade Mountains at The Dalles from November 18 to 22, 1921, with a 54-in (137-cm)-deep icy mass, blocking all road and rail traffic. The storm extended west almost to Portland and east to Walla Walla. Huge snow slides contributed to the blockade. Nine passenger trains were stalled. Total precipitation at The Dalles in the Gorge was 8.9 in (226 mm), and on nearby mountains as much as 13.03 in (331 mm) fell.

From January 6 to 9, 1942, the lower Willamette Valley in northwest Oregon and the lowlands of southwest Washington had a mixture of sleet and freezing rain that accumulated to build up an icy sheath 0.75 in (1.9 cm) thick and remained several days. Extensive forest damage resulted, as well as the usual downed wires, power outages, and hazardous traffic conditions. Damage on the North Head peninsula in southwest Washington alone was estimated at $5 million; fruit orchard losses in the Hood River Valley were placed at $6 million. Freezing rain occurred north to south over the entire Willamette Valley on January 7, an unusual occurrence.

The Walla Walla section of Washington and the Columbia Gorge experienced "the most damaging ice storm in recent years" on January 17–19, 1970. Breaks occurred in major transmission lines over the Cascades with ice accumulations up to 1.5 in (4 cm) reported. Tree damage amounting to $6 million occurred in the Hood River Valley of Oregon.

Weather Records
Glossary

Extremes of Cold — States

State	°*F*	°*C*	*Date*	*Location*
Alabama	−27	−33	1966, Jan. 30	New Market
Alaska	−80	−62	1971, Jan. 23	Prospect Creek Camp
Arizona	−40	−40	1971, Jan. 7	Hawley
Arkansas	−29	−34	1905, Feb. 13	Pond
California	−45	−43	1937, Jan. 20	Boca
Colorado	−60	−51	1979, Jan. 1*	Maybell
Connecticut	−32	−36	1943, Feb. 16	Falls Village
Delaware	−17	−27	1893, Jan. 17	Millsboro
D.C.	−15	−26	1899, Feb. 11	Washington
Florida	−2	−19	1899, Feb. 13	Tallahassee
Georgia	−17	−27	1940, Jan. 27	Calhoun (near)
Hawaii	14	10	1961, Jan. 2	Haleakale, Maui Is.
Idaho	−60	−51	1943, Jan. 18	Island Park Dam
Illinois	−35	−37	1930, Jan. 22	Mount Carroll
Indiana	−35	−37	1951, Feb. 2	Greensburg
Iowa	−47	−44	1912, Jan. 12	Washta
Kansas	−40	−40	1905, Feb. 13	Lebanon
Kentucky	−34	−37	1963, Jan. 28	Cythiana
Louisiana	−16	−27	1899, Feb. 13	Minden
Maine	−48	−44	1925, Jan. 19	Van Buren
Maryland	−40	−40	1912, Jan. 13	Oakland
Massachusetts	−34	−37	1957, Jan. 18	Birch Hill Dam
Michigan	−51	−46	1934, Feb. 9	Vanderbilt
Minnesota	−59	−51	1903, Feb. 16*	Pokegama Dam
Mississippi	−19	−28	1966, Jan. 30	Corinth (near)
Missouri	−40	−40	1905, Feb. 13	Warsaw
Montana	−70	−57	1954, Jan. 20	Rogers Pass
Nebraska	−47	−44	1899, Feb. 12	Camp Clarke
Nevada	−50	−46	1937, Jan. 8	San Jacinto
New Hampshire	−46	−43	1925, Jan. 28	Pittsburg
New Jersey	−34	−37	1904, Jan. 5	River Vale
New Mexico	−50	−46	1951, Feb. 1	Gavilan
New York	−52	−47	1979, Feb. 18*	Old Forge
North Carolina	−29	−34	1966, Jan. 30	Mt. Mitchell
North Dakota	−60	−51	1936, Feb. 15	Parshall
Ohio	−39	−39	1899, Feb. 10	Milligan
Oklahoma	−27	−33	1930, Jan. 18*	Watts
Oregon	−54	−48	1930, Jan. 18*	Seneca
Pennsylvania	−42	−41	1904, Jan. 5	Smethport
Rhode Island	−23	−31	1942, Jan. 11	Kingston

* Also on earlier dates at the same or other places in the state.

State	°*F*	°*C*	*Date*	*Location*
South Carolina	−20	−29	1977, Jan. 18	Longcreek (near)
South Dakota	−58	−50	1936, Feb. 17	McIntosh
Tennessee	−32	−36	1917, Dec. 30	Mountain City
Texas	−23	−31	1933, Feb. 8*	Seminole
Utah	−50	−46	1913, Jan. 5*	Strawberry Tunnel
Vermont	−50	−46	1933, Dec. 30	Bloomfield
Virginia	−29	−34	1899, Feb. 10	Monterey
Washington	−48	−44	1968, Dec. 30	Mazama & Winthrop
West Virginia	−37	−38	1917, Dec. 30	Lewisburg
Wisconsin	−54	−48	1922, Jan. 24	Danbury
Wyoming	−63	−53	1933, Feb. 9	Moran

Extremes of Heat — States

State	°*F*	°*C*	*Date*	*Location*
Alabama	112	44	1925, Sept. 5	Centerville
Alaska	100	38	1915, June 27	Fort Yukon
Arizona	127	53	1905, July 7*	Parker
Arkansas	120	49	1936, Aug. 10	Ozark
California	134	57	1913, July 10	Death Valley
Colorado	118	48	1888, July 11	Bennett
Connecticut	105	41	1926, July 22	Waterbury
Delaware	110	43	1930, July 21	Millsboro
D.C.	106	41	1930, July 20*	Washington
Florida	109	43	1931, June 29	Monticello
Georgia	112	44	1952, July 24	Louisville
Hawaii	100	38	1931, Apr. 27	Pahala
Idaho	118	48	1934, July 28	Orofino
Illinois	117	47	1954, July 14	East St. Louis
Indiana	116	47	1936, July 14	Collegeville
Iowa	118	48	1934, July 20	Keokuk
Kansas	121	49	1936, July 24*	Alton (near)
Kentucky	114	46	1930, July 28	Greensburg
Louisiana	114	46	1936, Aug. 10	Plain Dealing
Maine	105	41	1911, July 10*	North Bridgton
Maryland	109	43	1936, July 10*	Cumberland & Frederick
Massachusetts	107	42	1975, Aug. 2	Chester & New Bedford
Michigan	112	44	1936, July 13	Mio

* Also on earlier dates at the same or other places in the state.

State	°F	°C	*Date*	*Location*
Minnesota	114	46	1936, July 6*	Moorhead
Mississippi	115	46	1930, July 29	Holly Springs
Missouri	118	48	1954, July 14*	Warsaw & Union
Montana	117	47	1937, July 5	Medicine Lake
Nebraska	118	48	1936, July 24*	Minden
Nevada	122	50	1954, June 23*	Overton
New Hampshire	106	41	1911, July 4	Nashua
New Jersey	110	43	1936, July 10	Runyon
New Mexico	116	47	1934, July 14*	Orogrande
New York	108	42	1926, July 22	Troy
North Carolina	109	43	1954, Sept. 7*	Weldon
North Dakota	121	49	1936, July 6	Steele
Ohio	113	45	1934, July 21*	Gallipolis (near)
Oklahoma	120	49	1943, July 26*	Toshomingo
Oregon	119	48	1898, Aug. 10*	Pendleton
Pennsylvania	111	44	1936, July 10*	Phoenixville
Rhode Island	104	40	1075, Aug. 2	Providence
South Carolina	111	44	1954, June 28*	Camden
South Dakota	120	49	1936, July 5	Gann Valley
Tennessee	113	45	1930, Aug. 9*	Perryville
Texas	120	49	1936, Aug. 12	Seymour
Utah	116	47	1892, June 28	Saint George
Vermont	105	41	1911, July 4	Vernon
Virginia	110	43	1954, July 15	Balcony Falls
Washington	118	48	1961, Aug. 5*	Ice Harbor Dam
West Virginia	112	44	1936, July 10*	Martinsburg
Wisconsin	114	46	1936, July 13	Wisconsin Dells
Wyoming	114	46	1900, July 12	Basin

* Also on earlier dates at the same or other places in the state.

Extremes of Precipitation — States

State	*Greatest in 24 Hours*	*Greatest Monthly*	*Greatest Annual*	*Least Annual*
Alabama	20.33 in (516 mm) Axis 4/13/55	34.86 in (885 mm) Robertsdale 7/16	106.57 in (2707 mm) Mt. Vernon Barrack 1853	22.00 in (559 mm) Primrose Farm 1954
Alaska	14.84 in (377 mm) Little Port Walter 12/6/64	70.99 in (1803 mm) Mac Leod Harbor 11/76	332.29 in (8440 mm) Mac Leod Harbor 1976	1.61 in (41 mm) Barrow 1935
Arizona	11.40 in (291 mm) Workman Creek 1 9/4–5/70	16.95 in (431 mm) Crown King 8/51	58.92 in (1497 mm) Hawley Lake 1978	0.07 in (1.8 mm) Davis Dam 1956
Arkansas	12.00 in (305 mm) Arkadelphia 6/28/05	23.86 in (606 mm) El Dorado 12/31	98.55 in (2503 mm) Newhope 1957	19.11 in (485 mm) Index 1936
California	26.12 in (663 mm) Hoegees Camp 1/22–23/43	71.54 in (1815 mm) Helen Mine 1/09	153.54 in (3900 mm) Monumental 1909	0.00 in (0 mm) Death Valley 1929
Colorado	11.08 in (256 mm) Holly 6/17/65	23.28 in (591 mm) Ruby 2/1897	92.84 in (2358 mm) Ruby 1897	1.69 in (43 mm) Buena Vista 1939
Connecticut	12.12 in (308 mm) Hartford 8/18–19/55	27.70 in (704 mm) Torrington 2 8/55	78.53 in (1995 mm) Burlington Dam 1955	23.60 in (599 mm) Baltic 1965
Delaware	7.83 in (199 mm) Odessa 6/27/38	17. 69 in (449 mm) Bridgeville 8/67	72.75 in (1848 mm) Lewes 1948	21.38 in (543 mm) Dover 1965
District of Columbia	7.31 in (186 mm) 24th & M Streets 8/11/29	17.45 in (443 mm) 24th & M Streets 9/34	61.33 in (1558 mm) 24th & M Streets 1889	18.79 in (477 mm) Washington 1826
Florida	38.70 in (983 mm) Yankeetown 9/5/50	42.33 in (1075 mm) Ft. Lauderdale 10/65	112.43 in (2856 mm) Wewahitchka 1966	22.45 in (573 mm) Key West 1961
Georgia	18.00 in (457 mm) St. George 8/28/11	30.23 in (768 mm) Blakely 7/16	122.16 in (3103 mm) Flat Top 1959	17.14 in (435 mm) Swainsboro 1954
Hawaii	38.00 in (965 mm) Kilauea Plantation 1/24–25/56	107.00 in (2718 mm) Kukui, Maui 3/42	578.00 in (14,681 mm) Kukui, Maui, 1931	0.19 in (5 mm) Kawaihae 1953

State	Greatest in 24 Hours	Greatest Monthly	Greatest Annual	Least Annual
Idaho	7.17 in (182 mm) Rattlesnake Creek 11/23/09	28.23 in (717 mm) Roland 12/23	81.05 in (2059 mm) Roland 1933	2.09 in (53 mm) Grand View 1947
Illinois	16.54 in (420 mm) East St. Louis 6/14/57	20.03 in (509 mm) Monmouth 9/11	74.58 in (1894 mm) New Burnside 1950	16.59 in (421 mm) Keithsburg 1956
Indiana	10.50 in (267 mm) Princeton 8/6/05	21.39 in (543 mm) Evans Landing 1/37	97.38 in (2473 mm) Marengo 1890	18.67 in (474 mm) Brookville 1934
Iowa	16.70 in (424 mm) Decatur Co. 8/5–6/59	22.18 in (563 mm) Red Oak 6/67	74.50 in (1892 mm) Muscatine 1851	12.11 in (308 mm) Cherokee 1958
Kansas	12.59 in (320 mm) Burlington 5/31/41	24.56 in (624 mm) Ft. Scott 6/1845	65.87 in (1673 mm) Mound City 1951	4.77 in (121 mm) Johnson 1956
Kentucky	10.40 in (264 mm) Dunmor 6/28/60	22.97 in (583 mm) Earlington 1/37	79.68 in (2024 mm) Russellville 1950	14.51 in (368 mm) Jeremiah 1968
Louisiana	22.00 in (559 mm) Hackberry 8/29/62	37.99 in (965 mm) Lafayette 8/40	111.28 in (2827 mm) Morgan City 1946	26.44 in (672 mm) Shreveport 1936
Maine	8.05 in (204 mm) Brunswick 9/11/54	17.75 in (451 mm) Brunswick 11/1845	75.64 in (1921 mm) Brunswick 1845	23.06 in (586 mm) Machias 1930
Maryland	14.75 in (375 mm) Jewell 7/26/1897	20.35 in (517 mm) Leonardtown 7/45	72.59 in (1844 mm) Salisbury 1948	17.76 in (451 mm) Picardy 1930
Massachusetts	18.15 in (461 mm) Westfield 8/18–19/55	26.85 in (682 mm) Westfield 8/55	70.33 in (1786 mm) Westfield 1955	21.76 in (553 mm) Chatham Life Station 1965
Michigan	9.78 in (248 mm) Bloomingdale 9/1/14	16.24 in (413 mm) Battle Creek 6/1883	64.01 in (1626 mm) Adrian 1881	15.64 in (397 mm) Croswell 1936
Minnesota	10.75 in (273 mm) Mahnomen 7/20/09	16.52 in (420 mm) Alexandria 8/1900	51.53 in (1309 mm) Grand Meadow 1911	7.81 in (198 mm) Angus 1936
Mississippi	15.68 in (398 mm) Columbus 7/9/68	30.75 in (781 mm) Merrill 7/16	102.89 in (2613 mm) Beaumont 1961	25.97 in (660 mm) Yazoo City 1936

State	*Greatest in 24 Hours*	*Greatest Monthly*	*Greatest Annual*	*Least Annual*
Missouri	18.18 in (462 mm) Edgerton 7/20/65	25.54 in (649 mm) Joplin 5/43	92.77 in (2356 mm) Portageville 1957	16.14 in (410 mm) La Belle 1956
Montana	11.50 in (292 mm) Circle 6/20/21	16.79 in (426 mm) Circle 6/21	55.51 in (1410 mm) Summit 1953	2.97 in (75 mm) Belfry 1960
Nebraska	13.15 in (334 mm) York 7/8/50	20.00 in (508 mm) Tecumseh 6/1883	64.52 in (1639 mm) Omaha 1869	6.30 in (160 mm) Hull 1931
Nevada	7.40 in (188 mm) Lewers Ranch 3/18/07	33.03 in (839 mm) Mt. Rose 12/64	50.03 in (1499 mm) Mt. Rose 1969	Trace Hot Springs 1898
New Hampshire	10.38 in (264 mm) Mt. Washington 2/10–11/70	25.56 in (649 mm) Mt. Washington 2/69	130.14 in (3306 mm) Mt. Washington 1969	22.31 in (567 mm) Bethlehem 1930
New Jersey	14.81 in (376 mm) Tuckerton 8/19/39	25.98 in (660 mm) Paterson 9/1881	85.99 in (2184 mm) Paterson 1882	19.85 in (504 mm) Canton 1965
New Mexico	11.28 in (287 mm) Lake Maloya 5/19/55	16.21 in (412 mm) Portales 5/41	62.45 in (1586 mm) White Tail 1941	1.00 in (25mm) Hermanas 1910
New York	11.17 in (284 mm) NYC Central Park 10/9/03	25.27 in (642 mm) West Shokan 10/55	82.06 in (2084 mm) Wappingers Falls 1903	17.64 in (448 mm) Lewiston 1941
North Carolina	22.22 in (564 mm) Altapass 7/15/16	37.40 in (950 mm) Gorge 7/16	129.60 in (3291 mm) Rosman 1964	22.69 in (576 mm) Mount Airy 1930
North Dakota	7.70 in (196 mm) McKinney 6/15/1897	14.01 in (356 mm) Mohall 6/44	37.98 in (965 mm) Milnor 1944	4.02 in (102 mm) Parshall (near) 1934
Ohio	10.51 in (267 mm) Sandusky 7/12/66	16.13 in (410 mm) Deamos 7/1896	70.82 in (1799 mm) Little Mountain 1870	16.96 in (431 mm) Elyria 1963
Oklahoma	15.50 in (394 mm) Sapulpa 9/3–4/40	23.95 in (608 mm) Miami 5/43	84.47 in (2146 mm) Kiamichi Tower 1957	6.53 in (166 mm) Regnier 1956
Oregon	10.17 in (258 mm) Glenora 12/21/15	50.20 in (1271 mm) Glenora 11/09	168.88 in (4290 mm) Valsetz 1937	3.33 in (85 mm) Warmspring Reservoir 1939

State	*Greatest in 24 Hours*	*Greatest Monthly*	*Greatest Annual*	*Least Annual*
Pennsylvania	34.50 in (876 mm)* Smethport 7/17/42	23.66 in (601 mm) Mt. Pocono 8/55	81.64 in (2074 mm) Mt. Pocono 1952	15.71 in (399 mm) Breezewood 1965
Puerto Rico	23.00 in (584 mm) Adjuntas 8/8/1899	52.02 in (1321 mm) Jayoya 10/70	253.79 in (6446 mm) La Mina El Yunguc 1936	9.76 in (248 mm) Rio Jueyes 1967
Rhode Island	12.13 in (308 mm) Westerly 9/16–17/32	15.00 in (381 mm) Rocky Hill 8/55	65.91 in (1674 mm) Pawtucket 1888	24.08 in (612 mm) Block Island 1965
South Carolina	13.25 in (337 mm) Effingham 7/15/16	31.13 in (791 mm) Kingtree 7/16	101.65 in (2582 mm) Caesar's Head 1961	20.73 in (527 mm) Rock Hill 1954
South Dakota	8.00 in (203 mm) Elk Point 9/10/1900	18.61 in (473 mm) Deadwood 5/46	48.42 in (1230 mm) Deadwood 1946	2.89 in (48 mm) Ludlow 1936
Tennessee	11.00 in (279 mm) McMinnville 3/28/02	23.90 in (601 mm) McKenzie 1/37	114.88 in (2918 mm) Haw Knob 1957	25.23 in (641 mm) Halls 1941
Texas	38.20 in (970 mm)* Thrall 9/9–10/21	34.85 in (885 mm) McKinney 5/1881	109.38 in (2778 mm) Clarkville 1873	1.64 in (42 mm) Presidio 1956
Utah	6.00 in (152 mm) Bug Point 9/5/70	19.14 in (486 mm) Buckboard Flat 10/72	70.71 in (1796 mm) Alta 1975	1.34 in (34 mm) Myton 1974
Vermont	8.77 in (223 mm) Somerset 11/3–4/27	16.99 in (432 mm) Mays Mill 10/55	73.61 in (1870 mm) Mt. Mansfield 1969	22.98 in (584 mm) Burlington 1941
Virginia	27.00 in (686 mm)* Nelson County 8/20/69	23.88 in (607 mm) Big Meadows 8/55	81.78 in (2077 mm) Montebello 1972	12.52 in (318 mm) Moores Creek Dam 1941
Washington	12.00 in (305 mm) Quinault Ranger Station 1/21/35	57.04 in (1449 mm) Peterson's Ranch 12/33	184.56 in (4688 mm) Wynoochee Oxbow 1931	2.61 in (66 mm) Wahluke 1930
West Virginia	19.00 in (483 mm)* Rockport 7/18/1889	16.30 in (414 mm) Princeton 6/01	94.01 in (2388 mm) Romney 1948	9.50 in (241 mm) Upper Tract 1930

* Estimated from hydrologic bucket survey.

State	*Greatest in 24 Hours*	*Greatest Monthly*	*Greatest Annual*	*Least Annual*
Wisconsin	11.72 in (298 mm) Mellen 6/24/46	17.41 in (442 mm) Hayward 8/41	62.07 in (1577 mm) Embarrass 1884	12.00 in (305 mm) Plum Island 1937
Wyoming	5.50 in (140 mm) Dull Center 5/31/27	12.78 in (325 mm) Alva 5/62	55.46 in (1409 mm) Grassy Lake Dam 1945	1.28 in (33 mm) Lysite 1960

Snowstorms — States

State	*Greatest in 24 Hours*	*Greatest Single Storm*	*Greatest Month*	*Greatest Season*
Alabama	19.2 in (49 cm) Florence 12/31–1/1/64	19.5 in (50 cm) Florence 12/31–1/1/64	24.0 in (61 cm) Valley Head 1/40	25.0 in (64 cm) Valley Head 1939–40
Alaska	62.0 in (157 cm) Thompson Pass 12/29/55	175.0 in (446 cm) Thompson Pass 12/26–31/55	297.9 in (757 cm) Thompson Pass 2/53	974.5 in (2475 cm) Thompson Pass 1952–53
Arizona	38.0 in (97 cm) Heber R.S. 12/14/67	67.0 in (170 cm) Heber R.S. 12/13–16/67	104.8 in (266 cm) Flagstaff 1/49	226.7 in (576 cm) Hawley Lake 1967–68
Arkansas	25.0 in (64 cm) Corning 1/22/18	25.0 in (64 cm) Corning 1/22/18	48.0 in (122 cm) Calico Rock 1/18	61.0 in (155 cm) Hardy 1917–18
California	60.0 in (152 cm) Giant Forest 1/18–19/33	189.0 in (480 cm) Shasta Ski Bowl 2/13–19/59	390.0 in (991 cm) Tamarack 1/11	884.0 in (2245 cm) Tamarack 1906–07
Colorado	75.8 in (193 cm) Silver Lake 4/14–15/21	141.0 in (358 cm) Ruby 3/23–30/1899	249.0 in (632 cm) Ruby 3/1899	779.0 in (1979 cm) Ruby 1896–97
Connecticut	28.0 in (71 cm) New Haven 3/12/1888	50.0 in (127 cm) Middletown 3/11–14/1888	73.6 in (187 cm) Norfolk 3/56	177.4 in (451 cm) Norfolk 1955–56
Delaware	25.0 in (64 cm) Dover 2/19/79	25.0 in (64 cm) Dover 2/19/79	36.0 in (91 cm) Milford 2/1899	49.5 in (126 cm) Wilmington 1957–58

State	*Greatest in 24 Hours*	*Greatest Single Storm*	*Greatest Month*	*Greatest Season*
Florida	4.0 in (10 cm) Milton Exp. Station 3/6/54	4.0 in (10 cm) Milton 3/6/54	4.0 in (10 cm) Milton 3/54	4.0 in (10 cm) Milton 1953–54
Georgia	19.3 in (49 cm) Cedartown 3/2–3/42	19.3 in (49 cm) Cedartown 3/2–3/42	26.5 in (67 cm) Diamond 2/1895	39.0 in (99 cm) Diamond 1894–95
Idaho	38.0 in (97 cm) Sun Valley 2/11/59	60.0 in (152 cm) Roland W. Portal 12/25–27/37	143.8 in (365 cm) Burke 1/54	441.8 in (1122 cm) Roland W. Portal 1949–50
Illinois	36.0 in (91 cm) Astoria 2/27–28/1900	37.8 in (96 cm) Astoria 2/27–28/1900	47.0 in (119 cm) Astoria 2/1900	77.0 in (196 cm) Chicago 1969–70
Indiana	20.0 in (51 cm) La Porte 2/12/44*	37.0 in (94 cm) La Porte 2/14–19/58	59.8 in (152 cm) La Porte 2/58	122.3 in (311 cm) La Porte 1962–63
Iowa	21.0 in (53 cm) Sibley 2/18/62	30.8 in (78 cm) Rock Rapids 2/17–21/62	42.0 in (107 cm) Osage, Northwood 3/15	90.4 in (230 cm) Northwood 1908–09
Kansas	26.0 in (66 cm) Fort Scott 12/28–29/54	37.0 in (94 cm) Olathe 3/23–24/12	55.9 in (142 cm) Olathe 3/12	100.1 in (254 cm) Goodland 1979–80
Kentucky	18.0 in (46 cm) Bowling Green 3/9/60 and Celicia 11/2/66	27.0 in (69 cm) Bowling Green 3/7–11/60	46.5 in (118 cm) Benham 3/60	108.2 in (275 cm) Benham 1959–60
Louisiana	24.0 in (61 cm) Rayne 2/14–15/1895	24.0 in (61 cm) Rayne 2/14–15/1895	24.0 in (61 cm) Rayne 2/1895	24.0 in (61 cm) Rayne 1894–95
Maine	35.0 in (89 cm) Middle Dam 11/23/43	56.0 in (142 cm) Long Falls Dam 2/24–28/69	88.3 in (224 cm) Long Falls Dam 2/69	238.5 in (606 cm) Long Falls Dam 1968–69
Maryland	31.0 in (79 cm) Clear Spring 3/29/42	36.0 in (91 cm) Edgemont 3/29–30/42	58.0 in (147 cm) Oakland 1/1895	174.9 in (444 cm) Deer Park 1901–02

* Figures have been exceeded at mountaintop stations.

State	Greatest in 24 Hours	Greatest Single Storm	Greatest Month	Greatest Season
Massachusetts	28.2 in (72 cm) Blue Hill, Milton 2/24–25/69	47.0 in (119 cm) Peru 3/2–5/47	78.0 in (198 cm) Monroe 2/1893	162.0 in (411 cm) Monroe 1892–93
Michigan	27.0 in (69 cm) Dunbar 3/29/47 and Ishpemig 10/23/29	46.1 in (117 cm) Calumet 1/15–20/50	115.3 in (293 cm) Calumet 1/50	298.3 in (758 cm) Herman 1868–69
Minnesota	28.0 in (71 cm) Pigeon R. Bridge 4/4–5/33	35.2 in (89 cm) Duluth 12/5–8/50	66.4 in (169 cm) Collegeville 3/65	147.5 in (375 cm) Pigeon R. Bridge 1936–37
Mississippi	18.0 in (46 cm) Mt. Pleasant 12/23/63 and Tunica 12/23/63	18.0 in (46 cm) Mt. Pleasant 12/23/63	23.0 in (58 cm) Cleveland 1/66	25.2 in (64 cm) Senatobia 1967–68
Missouri	27.6 in (70 cm) Neosho 3/16–17/70	27.6 in (70 cm) Neosho 3/16–17/70	47.5 in (121 cm) Poplar Bluff 1/18	70.3 in (179 cm) Maryville 1911–12
Montana	44.0 in (112 cm) Summit 1/20/72	57.0 in (145 cm) Summit 1/19–21/72	131.1 in (333 cm) Summit 1/72	406.5 in (1032 cm) Kings Hill 1958–59
Nebraska	24.0 in (61 cm) Hickman 2/11/65	41.0 in (104 cm) Chadron 1/2–4/49	59.6 in (157 cm) Chadron 1/49	104.9 in (266 cm) Kimball 1958–59
Nevada	25.0 in (64 cm) Mt. Rose Resort 1/20/69	75.0 in (191 cm) Mt. Rose Resort 1/18–22/69	124.0 in (315 cm) Mt. Rose Resort 1/69	323.0 in (820 cm) Mt. Rose Resort 1968–69
New Hampshire	56.0 in (142 cm) Randolph 11/22–23/43	77.0 in (196 cm) Pinkham Notch 2/24–28/69	130.0 in (330 cm) Pinkham Notch 2/69*	323.0 in (820 cm) Pinkham Notch 1968–69
New Jersey	29.7 in (75 cm) Long Branch 12/26–27/47	34.0 in (87 cm) Cape May 2/11–14/1899	50.1 in (127 cm) Freehold 12/1880	108.1 in (275 cm) Culvers Lake 1915–16
New Mexico	30.0 in (76 cm) Sandia Crest 12/29/58	40.0 in (102 cm) Corona 12/14–16/59	144.0 in (366 cm) Anchor Mine 3/12	483.0 in (1227 cm) Anchor Mine 1911–12

State	*Greatest in 24 Hours*	*Greatest Single Storm*	*Greatest Month*	*Greatest Season*
New York	54.0 in (137 cm) Barnes Corner 1/9/76	69.0 in (175 cm) Watertown 1/18–22/40	192.0 in (488 cm) Bennett Bridge 1/78	466.9 in (1186 cm) Hooker 1976–77
North Carolina	31.0 in (79 cm) Nashville 3/2/27	31.0 in (79 cm) Nashville 3/2/27	56.5 in (144 cm) Boone 3/60	100.7 in (256 cm) Banner Elk 1959–60
North Dakota	24.0 in (61 cm) Lisbon 2/15/15 and Berthold Agency 2/25/30	35.0 in (89 cm) Lisbon 2/13–15/15	45.5 in (116 cm) Tagus 4/70	99.9 in (254 cm) Pembina 1906–07
Ohio	20.7 in (53 cm) Youngstown 11/24–25/50	36.3 in (91 cm) Steubenville 11/24–26/50	69.5 in (177 cm) Chardon 12/62	161.5 in (410 cm) Chardon 1959–60
Oklahoma	23.0 in (58 cm) Buffalo 2/21/71	36.0 in (91 cm) Buffalo 2/21–22/71	39.5 in (100 cm) Buffalo 2/71	87.3 in (222 cm) Beaver 1911–12
Oregon	37.0 in (94 cm) Crater Lake 1/17/51**	95.0 in (241 cm) Crater Lake 1/15–19/51	256.0 in (650 cm) Crater Lake 1/33	879.0 in (2233 cm) Crater Lake 1932–33
Pennsylvania	38.0 in (97 cm) Morgantown 3/20/58	50.0 in (127 cm) Morgantown 3/19–21/58	86.0 in (218 cm) Blue Knob 12/1890	225.0 in (572 cm) Blue Knob 1890–91
Rhode Island	34.0 in (86 cm) Foster 2/8–9/45	34.0 in (86 cm) Foster 2/8–9/45	62.0 in (157 cm) Foster 3/56	122.6 in (311 cm) Foster 1947–48
South Carolina	24.0 in (61 cm) Rimini 2/9–10/73	28.9 in (73 cm) Caesar's Head 2/15–16/69	33.9 in (86 cm) Caesar's Head 2/69	60.3 in (153 cm) Caesar's Head 1968–69
South Dakota	38.0 in (97 cm) Dumont 3/27/50	60.0 in (152 cm) Dumont 3/26–28/50	94.0 in (239 cm) Dumont 3/50	258.2 in (656 cm) Lead 1976–77
Tennessee	22.0 in (56 cm) Morristown 3/9/60	28.0 in (71 cm) Westbourne 2/19–21/60	39.0 in (99 cm) Mountain City 3/60	75.5 in (192 cm) Mountain City 1959–60

* Figures have been exceeded at mountaintop stations.
** Also on earlier dates at the same or other places in the state.

State	*Greatest in 24 Hours*	*Greatest Single Storm*	*Greatest Month*	*Greatest Season*
Texas	24.0 in (61 cm) Plainview 2/3–4/56	33.0 in (84 cm) Hale Center 2/2–4/56	36.0 in (91 cm) Hale Center 2/56	65.0 in (165 cm) Romero 1923–24
Utah	35.0 in (89 cm) Kanush 2/9/53	105.0 in (267 cm) Alta 1/24–30/65	168.0 in (427 cm) Alta 1/67	663.0 in (1684 cm) Alta 1951–52
Vermont	33.0 in (84 cm) St. Johnsbury 2/25/69	50.0 in (127 cm) Readsboro 3/2–6/47	75.0 in (191 cm)* Waitsfield 12/69	197.5 in (502 cm) Waitsfield 1970–71
Virginia	33.0 in (84 cm) Big Meadows 3/6/62	42.0 in (107 cm) Big Meadows 3/6–7/62	54.0 in (137 cm) Warrenton 2/1899	109.9 in (279 cm) Mountain Lake 1977–78
Washington	52.0 in (132 cm) Winthrop 1/21/35	129.0 in (328 cm) Laconia 2/24–26/10	363.0 in (922 cm) Paradise R.S. 1/25	1122 in (2850 cm) Paradise R.S. 1971–72
West Virginia	34.0 in (86 cm) Bayard 4/27–28/28	57.0 in (145 cm) Pickens 11/23–30/50	104.0 in (264 cm) Terra Alta 1/77	301.4 in (766 cm) Kumbrabow State Forest 1959–60
Wisconsin	26.0 in (66 cm) Neillsville 12/27/04	30.0 in (76 cm) Racine 2/19–20/1898	80.5 in (204 cm) Gurney 12/68	230.0 in (584 cm) Gurney 1968–69
Wyoming	34.0 in (86 cm) Bechler River 1/28/33	52.0 in (132 cm) Bechler River 1/15–19/37	188.5 in (479 cm) Bechler River 1/33	491.6 in (1249 cm) Bechler River 1921–22

* Figures have been exceeded at mountaintop stations.

Glossary

advection — horizontal movement of any atmospheric element; for example, air, moisture, or heat.

air mass — an extensive body of air whose horizontal distribution of temperature and moisture is nearly uniform.

airstream — a substantial body of air with the same characteristics flowing with the general circulation.

anticyclone — an atmospheric pressure system characterized by relatively high pressure at its center and winds blowing outward in clockwise fashion; also called a high-pressure area or, simply, a high.

atmospheric pressure — the weight per unit area of the total mass of air above a given point; also called barometric pressure.

chinook — a wind descending a mountain side and warming in the process by dynamic compression; characteristic of the slopes of the Rocky Mountains.

circulation — the flow pattern of moving air. The general circulation is the large-scale flow characteristic of the semipermanent pressure systems, while the secondary circulation occurs in the more temporary, migratory high- and low-pressure systems.

coastal storm — a cyclonic, low-pressure system moving along a coastal plain or just offshore; it causes north to northeast winds over the land; along the Atlantic seaboard it is called a northeaster.

cold front — the interface or transition zone between advancing cold air and retreating warm air.

condensation — the process whereby a substance changes from the vapor phase to the liquid or solid phase; the opposite of evaporation.

conduction — transmission of heat by direct contact through a material substance, as distinguished from convection, advection, and radiation.

convection — transfer of heat by movement of material bodies. In meteorology, convection refers particularly to the thermally induced, vertical motion of air.

convergence — a distribution of wind movement that results in a net inflow of air into an area such as a low-pressure area.

cooling degree days — number provides a means of estimating relative energy requirements for air conditioning. It is arrived at in a similar method as degree days. Base temperature is 65°F. The number of degrees by which the mean temperature for the day exceeds 65 is the number of cooling degree days for that date: 90° mean = 25 cooling degree days.

cyclogenesis — the process leading to the development of a new low-pressure system, or the intensification of a pre-existing one.

cyclone — an atmospheric pressure system characterized by relatively low pressure at its center and winds blowing inward in a counterclockwise fashion (in the northern hemisphere); also known as a low-pressure system, or, simply, a low.

deepening — the decrease of pressure at the center of a storm system.

degree days — would be better understood by the public if called "heating degrees for the day." Based on the premise that artificial heating is not needed at mean temperatures above 65°F (18°C). The mean temperature is half the sum of the day's maximum and minimum temperatures. Subtracting the mean temperature, say 30°F, from the base of 65, equals 35 degree days. They are totaled for the month and season to determine a relative value for heat consumption and costs.

depression — an area of low pressure.

dew — liquid water droplets or other objects caused by condensation of water vapor from the air as a result of radiation cooling.

dew point — the temperature at which a parcel of air reaches saturation as it is cooled at constant pressure.

discontinuity — a term employed in meteorology to describe the rapid variation of the gradient of an element, such as the rate of pressure or temperature change at a front.

disturbance — an area of low pressure attended by storm conditions.
divergence — a distribution of wind movement that results in a net outflow of air from an area such as a high-pressure system.
dry-bulb temperature — the ordinary temperature of the air as distinguished from the wet-bulb temperature.
equinoctial storm — a violent storm of wind occurring, in popular belief, at the time of the year when the sun crosses the equator; also known as a line storm by sailors.
evaporation — the change of a substance from the liquid to the vapor or gaseous stage; the opposite of condensation.
extra-tropical cyclone — an atmospheric disturbance that either originated outside the tropics, or, having left the tropics, has lost the characteristics of a tropical storm.
eye of the storm — a roughly circular area of comparatively light winds and fair weather found in the center of many tropical storms.
filling — the increase of pressure at the center of a storm system.
front — the interface or transition zone between air masses of different densities and characteristics.
frontogenesis — the process that leads to the formation of a front.
frost — ice crystals formed on grass or other objects by the sublimation of water vapor from the air at below-freezing temperatures.
glaze — a sheath of transparent ice resulting from an ice storm.
high — a pressure system characterized by relatively high pressure at the center; usually attended by fair weather.
instability — a condition of the atmosphere whereby a parcel of air given an initial vertical impulse will tend to continue to move upward.
inversion — a reversal of the normal decrease of temperature with increasing altitude; above the inversion line the temperature increases, or decreases less rapidly.
jet stream — a zone of relatively strong winds concentrated within a narrow zone in the upper atmosphere; usually the location of the maximum winds imbedded in the westerlies.
kiloPascal — a unit measure of atmospheric or barometric pressure; equals 0.295 inch or 10 millibars.

mean temperature — the average of any series of temperatures observed over a period of time. The mean daily temperature is the average of 24 hourly temperatures. For convenience, it is usually given as the average of the maximum and minimum temperatures for a 24-hour period.

millibar — a pressure unit employed mainly in meteorology equal to 1000 dynes per square centimeter; equals 0.0295 inch or 0.1 kiloPascal.

occluded front — a composition of two fronts produced when a cold front overtakes a warm front, forcing the warm air aloft.

overrunning — the ascent of warm air over relatively cool air; usually occurs in advance of a warm front.

polar air — an air mass conditioned over the tundra or snow-covered terrain of high latitudes.

polar front — a semipermanent discontinuity separating cold polar easterly winds and relatively warm westerly winds of the middle latitudes.

precipitation — products of condensation that fall as rain, snow, hail, or drizzle.

prefrontal squall line — an unstable line of turbulence preceding a cold front at some distance, often accompanied by showers or thunderstorms.

radiation — the transfer of energy through space without the agency of intervening matter.

recurvature — the poleward turning of the path of a tropical storm or hurricane; usually occurs when the center comes under the influence of the westerlies in subtropical latitudes.

relative humidity — the ratio of the actual amount of water vapor in a given volume of air to the amount that could be present if the air were saturated, temperature remaining constant. Commonly expressed as a percentage.

ridge — an elongated area of high barometric pressure.

saturation — condition of a parcel of air holding a maximum of water vapor; a 100 percent relative-humidity condition exists.

secondary cold front — a cold front that may form behind the primary cold front and carry an even colder flow of air.

secondary depression — an area of low pressure that forms in a trough to the south or east of the primary storm center.

semipermanent high or low — one of the relatively stationary and stable pressure and wind systems; for example, the Icelandic Low, the Bermuda High.

source region — an area of nearly uniform surface characteristics over which large bodies of air stagnate and acquire a more or less equal horizontal distribution of temperature and moisture.

squall line — a well-marked line of instability ahead of a cold front accompanied by strong gusty winds, turbulence, and often heavy showers.

stationary front — an interface zone between cold and warm air that exhibits little or no movement.

steering — the process whereby the direction of movement of surface pressure systems is influenced by the circulation aloft.

storm surge — an abnormal rise of the water level along a shore as a result, primarily, of wind flow in a storm; also called storm tide, storm wave, hurricane wave. Often the most damaging feature of a tropical storm.

sublimation — the transition of a substance from the solid phase directly to the vapor phase, or vice versa, without passing through the intermediate liquid phase.

subsidence — the descending motion aloft of a body of air, usually within an anticyclone; causes a spreading-out and warming of the lower layers of the atmosphere.

temperature-humidity index — popularly called the THI or the comfort index. Based on the following formula: THI = 0.4 (dry bulb °F + wet bulb °F) = 15. The dry bulb temperature is the current air temperature; the wet bulb temperature is arrived at by evaporating water on the bulb of a matching thermometer. If the dry bulb is 90°F and the wet bulb 70°F, the THI is: 0.4 (90 + 70), or 64 + 15 = 79. When the THI is above 75, some people are uncomfortable; above 80, everyone is miserable.

tendency — the local rate of change of a meteorological element; usually refers to a barometric rise or fall.

tropical air — an air mass conditioned over the warm surfaces of tropical seas or land.

tropical cyclone — a low-pressure area originating in the tropics, having a warm central core and often developing an eye.

trough — an elongated area of low pressure.

typhoon — the name applied in the western Pacific Ocean to severe tropical storms; their structures are similar to hurricanes.

warm front — the line of advancing warm air that is displacing cooler air at the surface.

warm sector — the portion of a cyclone containing warm air, usually located in the southeastern sector of the storm system.

waterspout — a funnel-shaped, tornadolike cloud complex that originates over a body of water.

water vapor — atmospheric moisture in the invisible gaseous phase.

wet-bulb temperature — the degree indicated by a ventilated thermometer arrived at by evaporating water on the bulb surface. Resort to prepared tables gives humidity figures.

whirlwind — a rapidly whirling, small-scale vortex of air, often seen on hot still days.

wind — air in motion, occurring naturally in the atmosphere and caused by a difference in densities of nearby air parcels; refers to air moving parallel with the surface of the ground.

wind chill factor — a figure derived from the attempted calculation of heat loss from exposed human skin through the combination of particular temperatures and air speeds and involves heat loss from four factors: conduction, convection, radiation, and evaporation. It is based on a formula calculated from Dr. Paul Siple's experimental observations at the Antarctica base Little America of the effect of wind and cold on exposed human flesh. A chart has been constructed for quick determination of the factor if one knows the wind speed and the current temperature.